THE NEW VARROA HANDBOOK

BERNHARD MOBUS & CLIVE de BRUYN

First published in April 1993 by Northern Bee Books, Scout Bottom Farm, Mytholmroyd. HX7 5JS.

Book designed by K Sutcliffe

Cover Design by Len Hutton

Printed by Arc & Throstle Press, Nanholme Mill, Shaw Wood Road, Todmorden. OL14 6DA

I.S.B.N. 0 - 907908 - 67 - 5

CONTENTS

1 - THE BIOLOGY OF VARROA - *Bernhard Mobus*
   1.1 Introduction   6
   1.2 Biology   8
         Public Life   12
         Longevity   14
         Reproduction   15
   1.3 Biology Update   21
   1.4 Damage through Parasitism   23
         To Worker Bees and Brood   23
         To Drone Brood and Drones   26
         To the Colony   27
   1.5 Population Dynamics of the Parasite   30
   1.6 Non-Computerabilia   34
   1.7 Methods of Detection   37
   1.8 Control Methods   42
         Biological Methods   43
         Genetic selection   43
         Varroa Bait Comb   46
         Artificial Swarm   47
         Brood stop   48
         Physical/Mechanical Methods of Control   48
         Heat   48
         Dusts   49
         Chemotherapy   50
         Gases and Fumes   51
         Chemical aerosols   53
         Dusts and Powders   56
         Liquid spray formulations   56
         Systemic varroacides   57
         Problems of Residue   59
         Problems of Resistance   59
   1.9 Drugs in Use, Trade Names and Chemical Names   61
   1.10 Bibliography   66

2 - TREATING VARROA IN FRANCE - *Bernhard Mobus*   68
   2.1 Table.1   75
   2.2 Table.2   76

3 - LEARNING TO LIVE WITH VARROA IN THE UK - *Clive de Bruyn*

3.1 Introduction	78
3.2 Natural History	81
3.3 What Effect will Varroa Have?	87
3.4 The Detection of Varroasis	89
3.5 World Movement of Bees	102
3.6 Spread in the U.K.	108
3.7 Medicaments	119
3.8 Biotechnical Control	134
3.9 Integrated Control	140
3.10 Treatments (Non Chemical)	143
3.11 Breeding Resistance	146
3.12 Research	154
3.13 Epilogue	158

THE BIOLOGY OF VARROA
Bernhard Mobus

1.1 INTRODUCTION

When the first colonies were lost to varroasis in Japan, very few beekeepers realised the impact this parasite would have on beekeeping all over the world. If Apimondia Congress Reports can be taken as an indication of the importance of a problem, then that of varroa can be used as a classic example. Not long ago only one paper dealt with the parasite and its effect on colonies of honeybees. Since then the number of reports presented at such meetings have snowballed, and strict selection procedures have to be applied to remove any trivial contributions or those which are simply duplications of other work.

As the parasite spread among the colonies of European countries, many opinions were voiced in the as yet unaffected states, that things would not be as bad as scare-mongers were painting them. Even the term 'parasite' was still being questioned at a time when proof, by means of electrophoresis of varroa serum had been provided in apicultural reports. Bee pathologists pointed out that varroasis is no disease, because one varroa mite on a bee does no harm, and that heavy losses of colonies might well have been caused by the poisons which had been used, or by other, especially viral diseases.

Alright, beekeepers are used to exaggeration. All reportage of 'doom and gloom' was therefore nothing new to beekeepers, and they quickly became too bored to listen after one or two lectures on the subject. I am afraid that apathy and indifference was one approach, but trivialisation of the problem, especially by scientists and expert advisers, was probably the more harmful one. Neither attitude made the problem go away. No wonder, that whenever varroa was discovered in a country, the beekeeping fraternity clamoured that it had been uninformed, even misled in the matter of varroasis.

At the time of writing, *Varroa jacobsoni* is still making its rapid progress across the North American continent and has gained a foothold in Britain. Thousands of beekeepers all over the world are, as yet, uninformed and will be helpless when faced with an enemy which can only thrive when traditional methods of good beekeeping are applied to maintain strong - and therefore - productive colonies. Such colonies provide varroa with a veritable paradise.

Even now we do not know everything about the mite, and any uncertainty allows the sowers of doubts to spread their gospel that in other countries the bees survive, even thrive, and produce more honey than ever before. Do not listen to their sweet words, because without your help our colonies, once invaded, are doomed to die after a few years. Just the same, pay also little attention to the scaremongers who claim that there is no life after *Varroa jacobsoni*. It is true that leave-alone beekeeping condemns all colonies to a certain death. On the other hand, a new, often scientific approach to beekeeping with varroa has made better honey yields a reality. But such changes and improvements came about only after bitter experiences, and they were helped by an open mind driven by a determination to learn and succeed - and a lot of hard work.

This book will try not to exaggerate or trivialise - except in one way. From now on the mite will be with us for a long time, and we might as well trivialise its upper case V and write about varroa instead. The book hopes to be completely honest, however frightening the picture appears to be which it paints. It will rely on collating and reporting scientific work from several countries and many scientists. No promise is made to supply a recipe for a 'cure' by 'magic bullets', which then can let us sit back and twiddle our thumbs. At the moment there is no such treatment. Any potions, gadgets or products advertised in the bee press, and promising instant salvation, should be treated with suspicion.

Learning to live with the parasitic mite involves changes in beekeeping practices and in all traditional ways of management. Even mental attitudes towards bees and beekeeping must change. Until the magic bullet is discovered, or until the bee itself can learn to live with its enemy, all beekeeping will involve a lot more work and expense, as well as a better understanding of the honeybee's biology. In the end we may actually get a bit more honey from our bees.

In compiling material for the book I have used information from many sources. Here I want to express my thanks to all who have fought and are still fighting varroa on all fronts and have shared their experiences generously with others in articles and lectures. It is impossible to name them all. Thanks.

1.2 THE BIOLOGY OF VARROA

Ten, twelve years ago the mite 'varroa' had been unknown, except, maybe, among acarologists. Since then it has become better known and, at the same time, infamous. Oudemans was the first scientist to describe and name the mite in 1904, and its full biological name still gives him credit: *Varroa jacobsoni* (Oudemans). At the time of its discovery it was just another new species of parasitic mite and was of little interest. After all, in the Far East the colonies of the native hive bees, the Indian bee *Apis cerana* did not suffer greatly from the parasite and had learned to live with it. The Indian bee is widely distributed throughout the Far East, including China and Japan, but has never been of significant economic importance. No European honeybee existed there until recently, and vast tracks of land separated the two species.

No one knows when the parasite came across its new victim, the European bee *Apis mellifera* (in its various races), for the first time. No doubt, the British Raj (in his various ranks) had tried keeping bees in India to feel 'at home', but early attempts had usually been unsuccessful. Maybe even then it had been varroa which was to blame for the demise of imported colonies. Modern times made people of many countries of the third world more aware of the economic value of honey production, and Asian and Indian beekeepers also began to think that 'the grass is greener on the other side of the fence'. They imported colonies and queens of the more productive European honeybee.

Since then varroa obtained a firm foothold on the new host, and somehow, the parasitic mite was brought back to Europe. It is a useless exercise to delve here into the historic details of its migration. Russia was probably one of the first European countries to suffer heavy losses from the inadvertent importation of the mite, and from there it spread rapidly throughout Eastern Europe. From there it also was taken to North Africa. A further, independent invasion may have occurred in West Germany, where it is thought that an importation of Indian bees, brought in for the purpose of genetic studies, transferred the varroa mite to colonies of European bees. All continents, apart from Australia and New Zealand, are affected, and in Europe the mite has been found in most countries by now. Ireland and Norway are probably exceptions; but for how long can these countries remain free?

At first little was known about the natural history of varroa, and even apicultural scientists, who were suddenly confronted with the mite, could find little information. When the parasite became a dangerous bee-pest, its biology was suddenly examined in detail by scientists of all countries. At Apimondia meetings the number of papers on the subject rose within 10 years from a single paper to over 50. Much more needs discovering, but in this Handbook we will try and assemble much of the available and relevant knowledge from all corners of the globe.

Introduction:
The Mite Varroa jacobsoni

Biologists classify everything, and *Varroa jacobsoni* also finds a place in the biological databank called taxonomy. Among the animals without backbones, the invertebrates, we have several classes with flattened, jointed appendages which are used for locomotion, defence and/or feeding, the arthropods. Crabs, centipedes, millipedes, spiders and all insects belong to these wee beasties - and the individual species are vast in number. All mites, ticks and spiders come into the class of arachnids, the spiderlike animals. They all have four pairs of walking legs which often do more than just enable the animal to walk. Arachnids are further characterised by having no antennae or compound eyes. The parasitic mites of the Order of acarina, some of them better known as mites and ticks (acari), have mouth parts which can pierce the skin of their host and draw blood or body fluids from it.

Some of the appendages of Arachnids also have a sexual function. For example, the transfer of sperm from the male to the female is not effected by direct copulatory union of the sexes (as we know it).

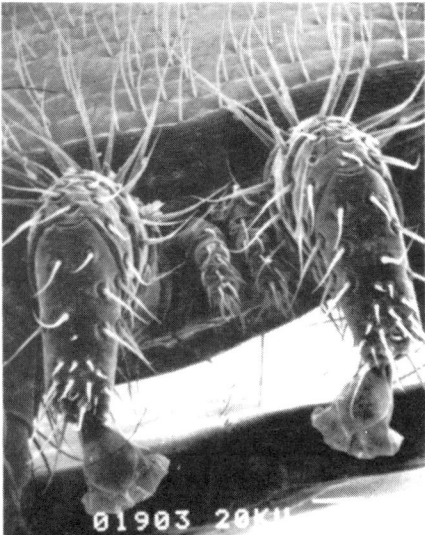

Fig 1 *Face to face with the enemy; Varroa Jacobsoni mouthparts between the front legs in close-up (bar = 50 µm).*

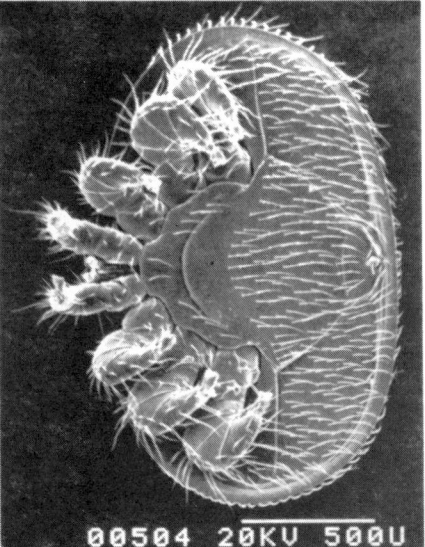

Fig 2 *Ventral view of a varroa mite showing legs and mouthparts. Abdominal plate seams are closed, indicating that the female is not in breeding condition. The crescent-shaped sexual opening stretching between the two hindmost legs can be seen clearly (bar = 500 µm)*

The New Varroa Handbook

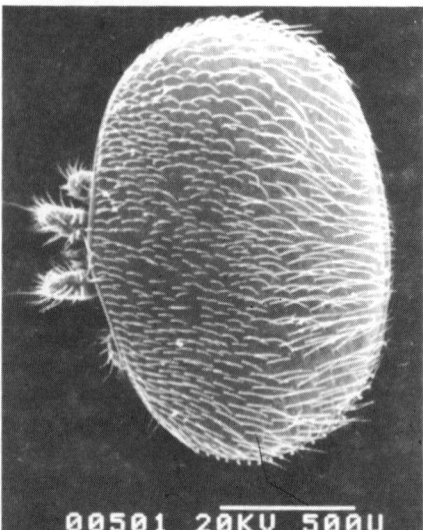

Fig 3 *Dorsal view of a varroa mite. Legs hardly protude beyond the body shield and there is no possibility of mistaking it for a bee louse (bar = 500 μm)*

Fig 4 *Varroa and the bee louse side by side at the same magnification.*

Instead, spermatocytes are ejected as a 'package' by the male from its sexual opening and are transferred into the sexual opening of the female by 'hand' (the 'hands' being modified mouth parts called pedipalps or chelicerae). The varroa male makes no exception - but we will not find a male specimen among the debris on the hive floor in order to examine him closely.

The bodies in the Acari are usually separated into two sections,

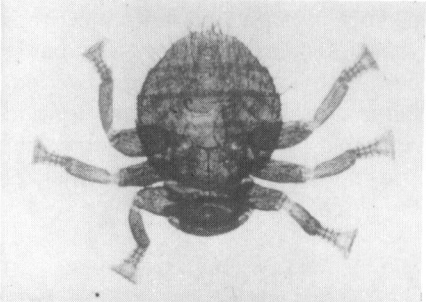

Fig 5 *Braula coeca, the bee louse. Three pairs of legs protrude beyond the body and have comb-like claws which make them cling to the body hair of bees.*

cephalothorax (also called gnathosoma) and the abdomen (idiosoma), and in the true ticks the abdomen is capable of considerable expansion in order to take in and hold the occasional large meal of blood. In varroa mites the body cavity can only hold a small amount of food at a time, because a hard body-shield encloses the varroa mite completely. The dorsal side of this carapace is smooth, though hairy, shows no segmentation and is incapable of expansion. It is of red-brown colour, oval in shape (1.2mm X 1.6mm) and slightly domed. The four pairs of legs hardly protrude from it. Short, stout hairs are distributed over the body shield, the abdominal segments and legs. The mouth parts of the female mite are adapted to pierce and draw blood from the soft skin of its victim, either larva, pupa or the adult bee. The last foot (tarsal) joint carries a sticky pad by means of which the mite can adhere to the bee's hairs. At point of death these pads release their hold and dead mites usually drop off the living host bee and to the floor of the hive.

10

Bernhard Mobus

To the beekeeper's eyes the difference between the harmless bee-louse, *Braula coeca*, and the dangerous varroa is small - unless he has very good eyesight. The mature adults of both are about the same size and brown colour. Often the individuals of both species adopt the same, safe place of refuge on the adult bee: on the thorax between the wings. Here the bee is least able to scratch itself or shake off its travelling companion. Very few beekeepers' eyes are sharp enough to tell the two apart, - never mind count the legs, and the two species will be confused unless they are magnified by lens or microscope.

Their life styles could not be more different. The bee louse is a 'commensal'* in the hive; it lives in close association with the honeybee, but the relationship between the two species is, although obligatory for the bee louse, neither parasitic* nor symbiotic*.

Just the same, when a queen carries too many *Braulae* on her body, this must amount to considerable irritation. Varroa, on the other hand, is a parasite, a blood sucking mite, which is certainly harmfull to its host, both adult bee and older bee brood. Apart from drawing the life blood of its host, the salival secretions in arachnoids contain protein-dissolving substances which predigest tissue or blood cells. Varroa seems to be no exception and much hidden damage is probably done to the host larva or bee in this way.

The bee-louse lays its eggs on the cappings of sealed honey comb during late spring and summer. Larvae emerge after a while and tunnel through the wax, causing unsightly white 'crazy paving' patterns in the cappings. The tunneling may lower the price of comb honey, but cause no harm to the honey or the bee. On the other hand, the reproduction of the parasitic mite is inescapably linked to bee brood within the host colony; it must take place in brood cells. The blood (more correctly called the haemolymph) of the developing larva (later pupa) is serving as the trigger for egg production and as nourishment for the growing generation of young mites. Furthermore, outside the brood cell, bee blood, drawn repeatedly from adult bees, sustains the adult varroa mite throughout its life.

> * Commensal: Different organisms living together in close harmony without infuencing each other.
> * Symbiotic: Association of dissimilar organisms to their mutual benefit.
> * Parasitism: Organism living in or on another species, its host, and obtaining food from it. Such an asociation may or may not be harmful to the host.

Fig. 6. *Adult varroa mite on workerbee.*

The New Varroa Handbook

Fig 7 *Adult bee with varroa mite nestling in the gap between thorax and abdomen.*

Public Life

We can divide the life cycle of the varroa mite into several convenient phases. In this booklet we start with the adult mite and its behaviour, with its activity and length of life 'in public'.

In a colony without sealed brood we will find only mature and, usually, fertilised female individuals. Male mites are small and immature when the young bee emerges from its cell, and they die after mating with the sisters.

The 'public life' of a varroa mite outside the brood cells begins when a young, adult mite emerges - together with its host bee, from the cell. The mite quickly climbs on a young worker or drone bee and, when it feels hungry, finds a gap between the overlapping abdominal segments of its host. Here it squeezes between the chitinous layers until it finds and bites through the soft, folded, intersegmental skin in order to draw the bee's blood for a good feed. With a full tummy, the young mite can live outside the brood cells for a few days before needing another feed. In the case of autumn-born varroa, the mites live on the adult bees of the winter cluster, sucking their blood at regular intervals to sustain themselves.

When winter is over and brood rearing has begun in the host colony, the surviving mites enter cells of worker brood (no drone brood in the first batches) in order to reproduce. A varroa mite enters a cell containing a larva just getting too big to lie comfortably curled in the base of the cell, or about 1- 2 days before the cell is sealed. This is between the 7th to 8th day (8th 9th day in drones) after the egg had been laid.

Drone cells are the preferred choice for the parasitic mite as soon as they become available. In the Indian bee the reproduction of varroa mites was only possible in drone cells, and chosing them became a biological necessity. We are uncertain about the trigger which makes varroa select drone cells. Size of cells is one possible factor, and taste, smell or some other distinctive drone brood pheromone could also act as trigger. The mite's long association with its original host - a different species with different 'smells', allows us to think that size of the cell or the grub is probably important. Some work actually supports this idea, another worker believes the attraction is due to the greater activity of nurse bees which are taking care of drone brood.

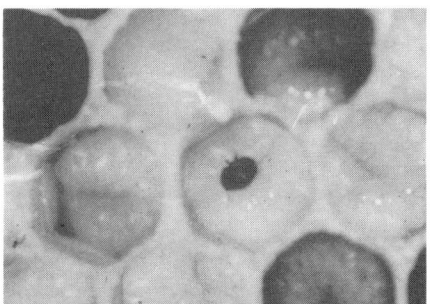

FIG 8 What the beekeeper must look out for: a fully coloured, adult varroa mite has entered a brood cell with a larva about 4 days after hatching

It seems that varroa females avoid a cell which is already occupied by another mite, at least as long as many brood cells are available. This selectiveness helps to reduce the damage done to an individual bee, and provides optimum conditions for the mite and her progeny at the same time. When the population pressure of mites is intense, or when too many mites are searching for too few larvae of the right age (brood nest contraction in autumn or when swarming conditions are initiated, more than one varroa will enter one cell. One report speaks of 21 mites found in one drone cell. This must be an exception; the demand for nourishment would effectively cripple and kill the host larva before it completes its development - as well as all parasites in the cell. Biologically it makes no sense.

Many adult varroa mites die during the winter period and before brood rearing begins in earnest. No hard and fast rule can be given. On the other hand, some regeneration is possible when brood rearing is initiated within the winter cluster. In such a case the losses and gains certainly are not balanced, and the overall reduction of the population of mites is considerable during cold winter months. In countries without a true winter period, and where pollen collection from early flowering plants stimulates early brood rearing, varroa regenerates relentlessly.

Some varroa die a natural death and drop from the bodies of their living host bees. They join the debris on the hive floor and can be detected. All the same, many more mites are lost when the host bee dies of natural causes and falls from the cluster. Up to 60% of the winter deaths of varroa may not be found or counted. They were probably feeding and hiding between the dying bee's abdominal segments. Natural and accidental deaths finally leave a spring population of variable size and variable

The New Varroa Handbook

age. On average - but just as in the bee world one should never think in such terms - the number of surviving mites is only a part of that which had been present during the previous autumn. Earlier observations in Europe suggested a ten-fold increase in the varroa population from spring count to spring count, but the figure needs revising, especially under different environmental conditions. To some extent, survival of winter mites also depends on strength and health of the colony, with an accellerated decline in numbers setting in a few weeks before the complete collapse of the host colony.

Longevity

Just as it does in the host bees, length of life varies considerably among the varroa mites. To begin with, we have short-lived summer varroa and long-lived winter mites. This is fairly obvious, but precise information on longevity needs many data. This creates a problem; because the normal life of individual varroa (marked?) can only be studied within the colony in which every inspection or investigation is bound to cause a disturbance. Investigating the mite's behaviour in minute observation hives in an incubator is also an unnatural situation for host and parasite, and some observations must remain suspect.

Schulz (1984) found that 78% of the reproductives entered a cell only once, with 22% reproducing twice. More recent investigations have shown that summer mites can, exceptionally, enter brood cells up to seven times (not necessarily for successful reproduction every time!), with the average probably lying between 3 - 4 reproductive cycles. With cells remaining sealed for around 12 days (worker) or 14 days (drone), the life expectancy of a mite can be calculated to reach 50 days on average, with the exceptional Methusalah among the summer mites attaining the ripe old

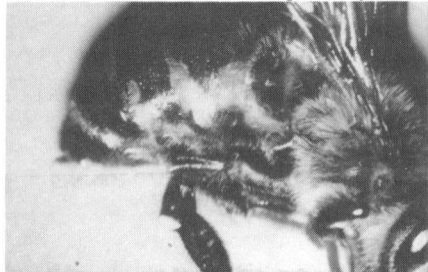

FIG 9 *Varroa mite sucking haemolymph, partially hidden between abdominal tergites of worker bee.*

age of 3 months. Non-breeding winter mites can live longer, up to 9 months, but may well have fewer reproductive cycles in their old age by the following spring. In countries with early falls and long winters, the population losses can be severe for the parasite.

Behaviour of Adult Mites

During summer months all young varroa mites which have just emerged from the host cell and are not yet ready to enter brood cells must 'live' on host bees. It is possibly a maturing process. Although thousands of bees are milling around, the mites have certain preferences even then. Young bees of the brood nest are favourite hosts, while the old foragers are least likely to be carriers. In experiments in which the whole brood nest, together with all nurse bees, was set aside and replaced with a new hive (with queen), the foragers returning to the old stand carried very few mites. Only 0.5 - 0.9% of the total original population mite was found in bait combs offered to the new artificial swarm (RITTER 1987). Drones, on the other hand, are more likely to carry varroa mites, and up to 23% of those leaving the hive may be infested.

Varroa mites outside the brood cells must draw the blood of adult bees. The mite's mouthparts are not strong enough to pierce the hard chitin of the exosceleton, and it therefore slides between the overlapping segments of the abdomen. Varroa's body shape allows it to penetrate to the soft inter-segmental skin where the sharp chelicerae cut a hole and tap the haemolymph. A mite can draw about 0.1 mg of blood in two hours, and that lasts for some time. No work has been done on the frequency of feeding visits, but the length of the mite's survival away from bees can give us a clue. Varroa mites have survived for up to 7 days outwith the colony, and so we can assume, that in an emergency one feed in the hive can also provide sufficient energy for a similar period. Mites which were still in a sealed cell when the colony died, lived for 3 weeks, no doubt getting sustenance from the dead pupal host. On the other hand, the transfer of bacilli from parasite to host larva was successful when infected mites remained only 24 hours in a fresh cell with a healthy larva.

FIG 10 Varroa female mite on the prowl. The forelegs are questing and are used like antennae to probe the environmen. One of the peritremes, snorkel-like, albeit closed-ended, extentions to the breathing apparatus can be seen above the third leg.

FIG 11 Ventral view of three varroa mites. Upper left: as found on adult worker bees; upper right; in a brood cell after a meal of blood and with developing egg; centre: an egg is protruding from the sexual opening.

The Reproductive Phase: In the Cell

Schultz (1984) reported that 44% of the young mites had slipped into cells (when these were available in adequate numbers) within 6 days. The percentage had risen to 69% after 12 days, and 90% of all young varroa were to be found on their larval victims after 24 days. Ritter reported that young mites remain on adult bees for 5 - 13 days before entering brood cells. These figures do not clash with reports that 85 - 90% of adult varroa are to be found in brood cells during summer, but they do imply, that most mites found outwith a brood cell are of the younger age group, while mother mites quickly return to another cell.

When a varroa mite enters an open brood cell in order to reproduce, it is soon cut off from the outside world by the capping of wax. This is later reinforced by the cocoon. Cappings over brood are not pure wax, but contain a lot of dross and detritus, including old cappings. That makes them permeable for the exchange of air and waste gases, so that larvae and pupae, as well as the parasites, can breath. Shortly after the cell has been sealed by worker bees, the bee larva stretches out, defecates and spins its cocoon. Then begins a period of rest and pupation. Infantidis, a Greek observer, reported

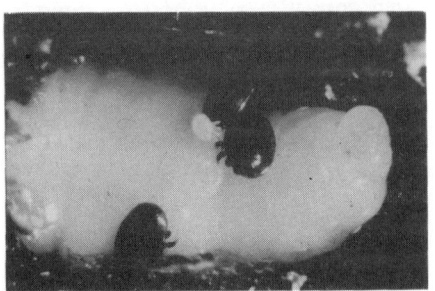

FIG 13 *Three mature varroa mites and one protonymph feeding on pre-pupa. A multiple invasion of a brood cell.*

that mites entering a cell were first trapped in surplus brood food at the base of the cell. They were said to remain submerged in the sea of brood food, drawing a breath of air through peritremes. Peritremes are small, external extensions to the tracheal system and are, in small mites, closed-ended (see **Fig.10**). They do act as oxygen exchangers, but cannot be used as snorkels as reported in some articles. When the larva had consumed the food, they liberated the mite. Any mites not released from their 'liquid prison' were said to die there. Normally, though, nurse bees do not mass-provision older worker larvae, and any fresh supplies given are quickly consumed. Liquid prison conditions may well have been due to experimental circumstances, while in a normal hive the losses due to 'drowning' are smaller.

On entering a cell, the varroa female begins to feed, piercing the skin and sucking the blood of the larva. Larval blood, or better, one of its components (juvenile hormone), stimulates the development of an egg cell and the

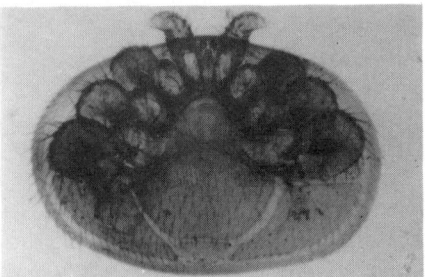

FIG 12 *Ventral view of a gravid varroa mite. Abdominal segments are expanding under internal pressure from developing egg.*

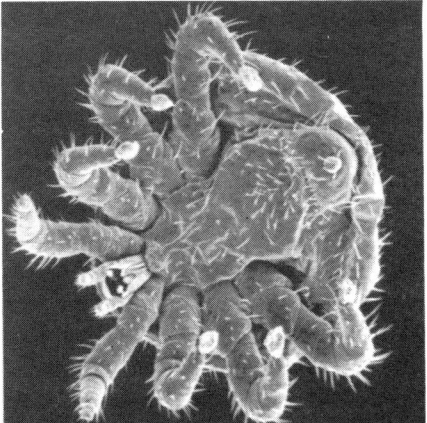

FIG 14 *Ventral view of the deuteronymph of varroa (bar = 500 μm)*

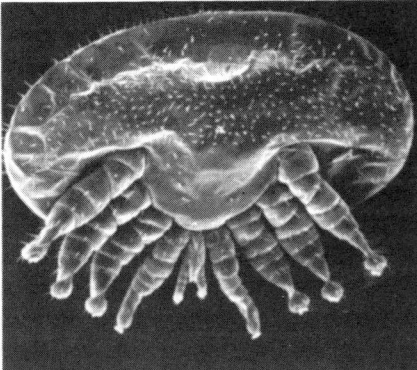

FIG 15 *Ventral view of protonymph of varroa (bar = 500 μm)*

maturation of a sperm-cell in the mite. This does not happen in all cases, and it has been suggested that about 10 - 20% of all varroa mites can not reproduce on entering a cell. The blood of adult bees in the open has no such stimulating effect on the ovaries of mites. The egg developing in the gravid mother mite now forces the ventral abdominal plates wide apart. After a minimum interval of 60 hrs, the first, by then fertilised, egg is laid, about two and a half days after the mite had its first meal.

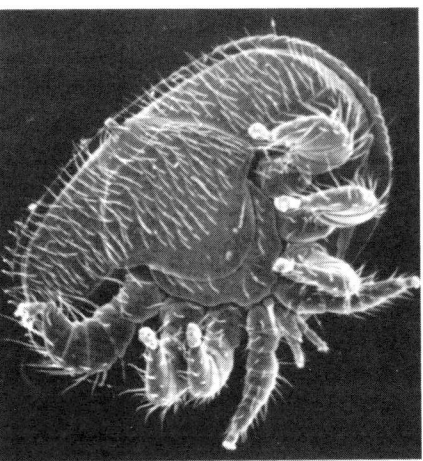

FIG 16 *Ventral view of the mature adult mite (bar = 500 μm)*

The first intake of haemolymph from the larva is sufficient to allow two eggs to grow, and the second egg is laid about 30 - 36 hours after the first. Lack of renewed stimulation prevents the development of a spermatocyte, and the second egg laid is usually unfertilised. Just as in bees, an unfertilised egg grows to become a male mite by parthenogenesis. The mother-mite then feeds again from the, by now, pupa and continues to lay several more, fertilised eggs at about 30 hour intervals. In a worker cell the maximum number of eggs is five, in drone cells up to seven eggs may be laid. They are usually deposited on the cell wall, not on the larva itself.

Varroa; Egg to Mature Mite

Incubation time for varroa eggs is - again roughly, 24 hours at normal brood nest temperatures. The freshly-born varroa nymph, called a protonymph at this stage, is quite capable of piercing the soft integument of the host larva and

The New Varroa Handbook

drawing its first feed of blood. Passing through several stages of development, a female deuteronymph matures into an adult within 7.5 days - or 180 hrs after the egg had been laid. The male nymph born from the second egg grows more rapidly and is sexually mature after 5.5 days - or 132 hrs. Adding now the period of growth of the egg (60 hrs to first egg, 30 hrs interval after that), we come to a total of 10 days (60 + 180 = 240 hrs) for female mites, and 222hrs or 9.25 days for the male - counting from the point when mother varroa had

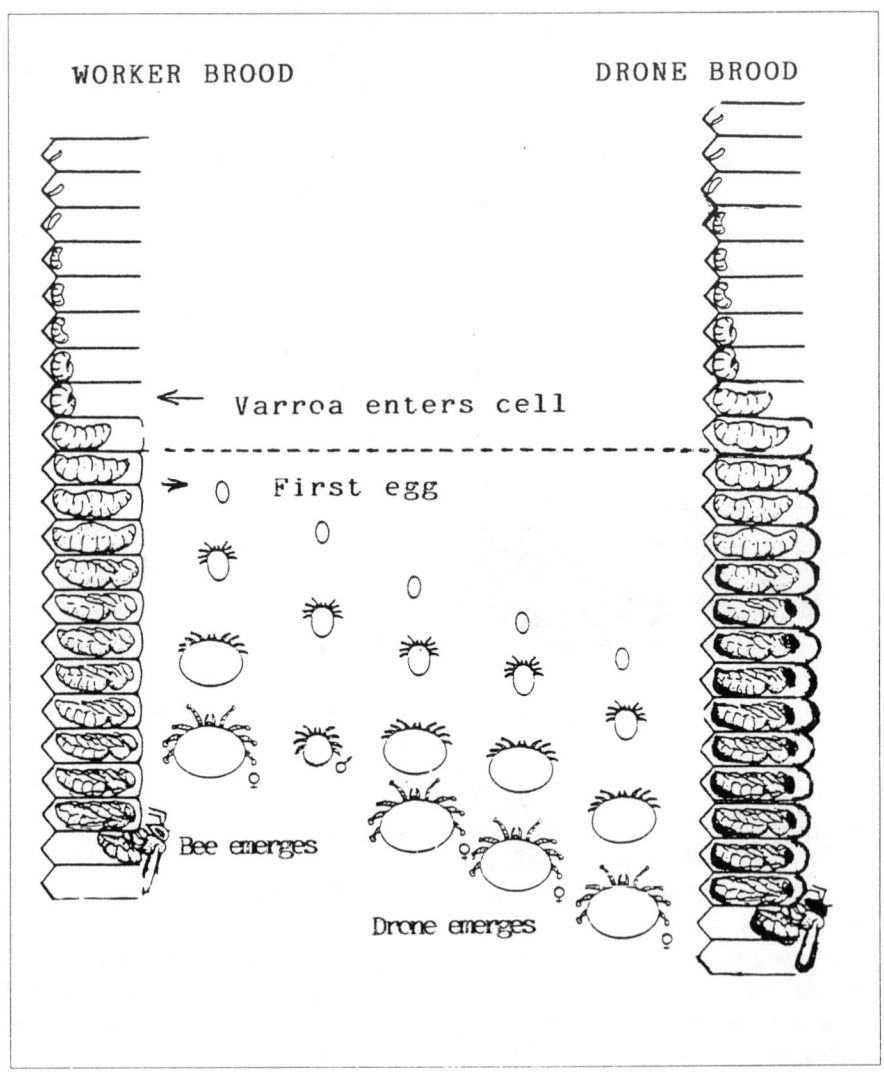

With kind permisson of the publishers of ADIZ (Allgemeine Deutsche Imkerzeitung)

its first feed. Before his sister is mature, the male, though still immature, is capable of inseminating his half-sister. Inbreeding is the rule, and has been for thousands of years. It has not led to a degeneration of vigour in varroa.

Third and consequent eggs are usually fertilised and will produce other female nymphs. The time passing before these eggs are laid adds 120 hrs, 150 hrs and 180 hrs respectively to the development of female nymphs (180 hrs), and so the third and subsequent mites (all female) will become mature 12.5; 13.75 and 15 days after the cell had been sealed.

Only when the mother mite takes its first feed well before the worker cell is sealed, and when the cell remains sealed for 12 days, can the second female become a mature mite. In worker cells all subsequent nymphs have 'missed the boat'; they remain immature, pale-coloured and will die when the young bee emerges. This reduces the 'productivity' of worker cells considerably (in terms of young mites per cell, not in young mites per mother), and the 'reproductive factor' rarely exceeds 1.3 per cycle, and that only when the development is slowed up when brood nest temperatures drop slightly in late summer.

In drone cells more mites can reach maturity. Drone brood remains sealed for 14 days, and up to four mites, three females and one male, can attain sexual maturity. On the whole, we can expect an average rate of reproduction in drone cells of better than 2.7, and the rate of the parasite's population expansion increases dramatically as soon as drone cells become available in spring. The earlier the rearing of drone brood begins, and the longer external conditions let colonies maintain patches of drone brood, the greater are the mite's chances for the development. Taking the preference for drone cells, their greater 'productivity', but also their seasonally variable, and usually smaller number into consideration, the overall reproductive rate will increase from a factor of probably less than 1.0 per cycle of worker brood in the first days of spring, to about 1.75 as soon as drone

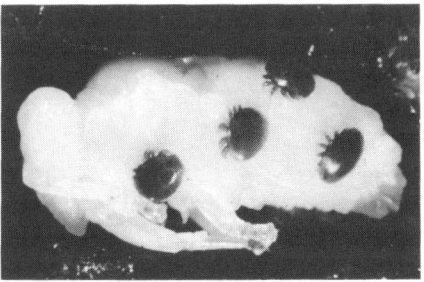

FIG 17 *Another way of detecting a varroa infection: fully mature mites can easily be seen on pink to purple eyed drone pupae*

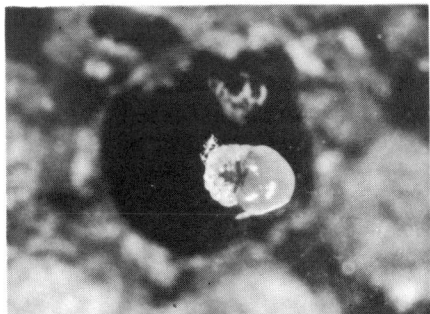

FIG 18 *Varroa nymph left inside a brood cell after emergence of worker bee. Immature stages of the parasite will perish.*

brood becomes available. Very strong colonies of highly prolific strains of bees in a district of mild winters are therefore the first to reach 'critical mass'.

After 11 - 12 days of development and pupation, the young bee emerges together with the mother-mite and its brood. The old mite can enter another cell and repeat the process of reproduction. The intervals between emergence and re-entering the next cell must be, on average, a short one for the sexually active individuals of the summer generations. However, the young mites need some time to mature before feeling the urge to reproduce.

N.B. This scheme of egg laying and the development of varroa in brood cells of bees, either drone or worker, was based on the observations made by Infantides in Greece, and many computer simulations of population dynamics of varroa have been based on them. In Germany another worker at the Beekeeping Institute of Freiburg, Miss Rehm had made more precise studies. Her work shows that, in biology at least, nothing is so thoroughly investigated that findings can be considered as final. The article which reported the new facts in **"The Beekeepers Quarterly"** can be found under **"Varroa Updated"**.

Bernhard Mobus

VARROA UPDATED

New information shows that some facts given in the Varroa Handbook must be updated. This information was gleaned by BM when he visited Dr. Ritter in Freiburg and ran into a Conference of the scientists who work at German Bee Research Institutes.

Among the lectures Miss Rehm overturned our 'knowledge' of the development of the nymphal stages of varroa, which is so well known and continuously cited ever since Dr. Ifantides published the result of his work. This shows once again that, in Biology especially, nothing is investigated so thoroughly that findings cannot be considered as final.

Brehm's research was done far more carefully, and varroa eggs were removed from cells infested with adult varroa mother mites. Transferred singly into new, non-infested cells, this left no doubt about the times of development of the mite via the protonymph, the deuteronymph into an adult mite capable of facing the world when the young bee emerges from its cell. The following facts must now be weighed up as influential in the population dynamics of mite infestations of colonies.

a) The development period for a female mite is 2 days shorter than the previously stated number of days. Rehm's work gives us 5.3 - 5.7 days from egg to adult.

b) The male varroa appears to grow at about the same speed as the female mite, namely 5.7 - 6 days.

c) The first egg often develops into a male mite. (Other workers attending the Conference suggested that this aspect should be repeated at different times of the year. It therefore seems that the incidence of male/female individuals may vary with the season.

The consequences of Miss Rehm's research must upset many scientists who dreamed of selecting for a bee capable of living with varroa without collapse due to a shortened period of development.

Her work means that under favourable circumstances, such as an early start to egg laying by the mother mite, as many as three female mites could mature in a worker cell. Even more mites can develop in a drone cell. All graphs and computer analyses of population dynamics of varroa populations must be updated in the light of this knowledge, and the picture which emerges is worse than originally conceived in the light of Infantides. Even the idea, that the mites cannot complete their development in a race or strain of bees which has a shortened capped period (Cape bees), must be discarded. The listing of the mite's development (**Table 1A** or **Fig. 2A**) shows that two female varroa mites could develop in worker cells of Cape bees which has a developing period of

19.5 days from the egg. It must also be clear that other factors seem to operate against the explosive, destructive growth of parasitic populations which could, theoretically at least, wipe out honeybee colonies in a very short time.

The level of juvenile hormone in bees appears to be of greater importance in determining the development of first eggs inside the mother mite than had been assumed until now. Lack of this hormone must be the factor which operates when a varroa mite enters a cell and lays no egg at all, or when the first egg is produced after a delay of several days. Early work on Indian bees gave the clue that the levels of juvenile hormone in worker bees of that race is only one quarter of that found in the blood of workers of the European (A.mellifera) races. (Hanel, 1985)

Since then, extensive work on the juvenile hormone titre in the haemolymph of worker honeybees has shown that it is subject to considerable variation during the life of a bee, as well as to seasonal fluctuations. Winter bees, and this includes bees who, from July/ August onwards, are not rearing brood and are just turning into winter bees, have a low hormone level. Active summer bees have a high level, and so have nurse bees in colonies with large brood nests where many larvae are making heavy demands on the glandular metabolism of nursing bees.

These researchers are now becoming convinced, that a low hormone titre in the adult bee's blood makes young varroa mites outside the cells mature more slowly and less 'efficient' in starting eggs production in worker cells - or drone cells for that matter.

It seems, that colonies of honeybees which maintain a small brood nest, yet are populous because of the adult bee's longevity, appear to withstand the onslaught of varroa far better than colonies of prolific races. In the first kind the ratio of potential nurse bees to brood is far greater. Not all bees are fully stretched by nursing duties. The life of an adult worker bee is, on the whole, a long one with a low hormone titre. Among the prolific races, on the other hand, or the highly bred 'hybrid' strains, we find that colonies have comparatively large brood nests, that each young bee has a lot of brood on its hands and must endure an exhausting nursing period. As a direct consequence of the wear and tear on glands, organs and tissues, such busy bees have a very short life. The juvenile hormone in the blood of the bees in the midst of a large brood nest is running at high levels.

Bernhard Mobus

DAMAGE THROUGH VARROA PARASITES 1.4

Varroa Damage to Individual Bees

It must be obvious to all that in small, short-lived insects the withdrawal of blood cannot remain without consequence. In varroasis of honeybees the parasite attacks along two fronts: adult bees are the victims of varroa mites outwith the cell, and they have to support the parasitic population, however large or small, throughout the winter months. Brood, mainly in the sealed stages, is the source of food for the actively reproducing females as well as their progeny. Bee brood is passing through several stages of development and feeding has stopped before the capping is constructed. Any growth and development of the mite and its brood must therefore be entirely at the expense of larval/pupal tissue, be it a worker, drone or a queen. Although beekeepers believe that 'only' blood is drawn, this is not the full story. The whole bee's body suffers, not only from the loss of nutrients, but also from the injection of the mite's salival secretions or the transfer

FIG 19 *Adult worker bees from a colony in Florida. All are visibly damaged by varroa mites.*

of disease agents. The amount of damage is in direct relation to the number of mites, both the matures and their brood, which tax the hapless victim. In the end, life, as seen in terms of performance, behaviour, health and longevity, must be affected by the 'loss of blood'.

It is far easier to follow a marked bee from 'cradle to grave', than to keep one's eye on a lot of adult bees of uncertain age within the colony. The latter investigation procedure can, at best, establish only 'guestimates' of the effect of parasitism on the host population. At best, it can produce averages only and questionable results. It is therefore better if we look first at the effect of an attack by one or several varroa mites on an individual bee 'in the cradle', the larva growing into a pupa and becoming finally a bee. Varroa mites enter a cell when the larva is nearly fully grown and when provisioning ceases. Early experiments assessed the damage as average weight loss at point of emergence, others delved more deeply and reported loss of protein in the emerging bee. Quoting these simple facts is not enough.

One thorough study repeated such work, but also observed the life of marked bees in later life. Most of the figures cited in this article come from that work (Schneider et al, 1988). This proved once more that unparasitised bees were 10% heavier than workers who had emerged with 3 or less fully coloured, mature mites. Bees leaving their cell with 4 or more mature mites were 22% lighter than healthy ones.

Nowadays we need more information and scientists look at other aspects of the damaged bee's life and its physiology. Since the early days of varroasis, the protein losses through parasitism in pupae and emerging bees have been well substantiated by several workers. An interesting fact emerged in one study: although the overall protein content of pupae showed losses, the protein concentration actually increased considerably in the haemolymph after mites had drawn blood. It is not quite proven whether the higher concentration is due to dehydration only, or due to the dissolution of tissue under the influence of proteolytic enzymes in the saliva of mother mites and their growing brood.

Since then, one other study has shown that the spectrum of blood cells changes dramatically when a bee is parasitised. Many of the blood cells, such as oenocytes, serve a specific function and any change from a healthy picture must render the bee more vulnerable to disease and rapid ageing.

Length of adult life is therefore greatly reduced for any worker bee which had to support a female mite and its young. Again, the effect is greater the more female mites had entered the cell in the first place and had reproduced successfully. A bee's life was halved, on average, when it had been the victim of one mite and its progeny. When two varroa entered a cell, the bee lived for about 17 days only. When more mites dived into a cell before 'closing time', that bee was so badly damaged that it lived only 9 days - if it did not emerge crippled. A worker of the two last categories can, of course, never become a nurse bee or a wax-producer.

Indeed, newest work (Schneider et al, 1988) has shown that worker bees* of Group 2 (bees emerging with more than 3 fully coloured, mature varroa mites) rarely left the hive at all. The same investigation produced the following results: worker bees emerging from unparasitised cells (Group 3) weighed 114 mg at point of emergence, while bees emerging with up to 3 mites (Group 1) were 10% lighter at 103 mg average. Young bees emerging in company of more than 3 parasites (Group 2) weighed, on average, only 89 mg, or 78% of the weight of a healthy bee.

* The study included drones, classified in the same way: D1; D2; D3, see later.

The same investigation then investigated the the health status of the secretory acini (berry-shaped cell clusters) of the brood food glands. These glands are vital for the future generations and the existence of the colony. Acini were greatly reduced in size in individuals of Group 1. Not even the consumption of pollen during first days of its life could fully compensate for the bee's loss of protein. Few bees of Group 2 ever reached the age of 16 days, and no measurement of brood food glands was attempted.

There should be no need to spell out in fine detail the full implication of the damage to the hypopharyngeal glands and the lack of brood food or the drop in its quality. Just the same, it must be done here in order to make it crystal-clear to all beekeepers. One mite can harm only one baby bee - at a time. That colony remains healthy. But when the parasitic population is increasing in leaps and bounds, there comes a time when the number of damaged young bees grows out of proportion. Yet they should become the nurse bees which should, in turn, raise future generations which, in the autumn should raise the long-lived winter bees. When the brood food glands of thousands of young bees are damaged, the damage to the colony becomes a chain reaction.

The lack of brood food, accompanied by a deterioration of its quality, will finally damage even brood which is not parasitised. Young larvae in open cells can suddenly grow too weak to resist any bugs and spores which are normally present - but rarely cause damage, begin to develop. First signs of brood damage may appear as an increased level of chalk brood. In advanced cases the brood nest presents a picture as if AFB has struck; when too many varroa mites enter a cell, some larvae die after the cells had been capped. Yet the true AFB bug may not be present. When the quality - as well as the quantity - of brood food diminishes, open brood may be so badly stressed that it dies before reaching sealing stage. Hungry larvae crawl to the front of the cell, the brood nest looking like a bad case of EFB.

Back now to the previous investigation by Schneider et al. In the course of that work, all worker bees, healthy or damaged, were colour-coded at point of emergence, and the flight behaviour was observed by means of a long tunnel entrance. Patient observations showed that very few bees of Group 2 made any flights at all. Healthy bees of Group 3 and those of Group 1 (emerging with fewer than 3 mites) made flights of similar frequency and duration between day 5 to 12. In older bees (13 - 20 days) the damage showed: the parasitised workers (Group 1) reached only 26% of total flying time when compared with healthy workers. Performance of this group 'improved' with age, and after 20 days their total flying time was only 18% behind that of healthy bees. We must accept this last result with caution. By that time the bees which had suffered most and had recovered least had - literally - dropped by the wayside. Only the least damaged bees of this group had stayed the course and made flights of nearly normal frequency and duration.

Damage to Drones

Schneider's study also investigated the damage to drones, both brood and adults. In drones the damage is a hidden one. The drone larva/pupa is larger and has greater reserves of fat and protein and can 'support' more mites. One mite entering a drone cell will therefore not have the same impact on the adult drone as on a worker, even though more young mites reach maturity. Just the same, the absolute weight loss per drone in Group D1 is of similar dimension and showed a 9mg weight loss per drone (in workers 11mg/bee). Proportionally to body weight this is far less than in worker bees. Drones of the heavily parasitised Group D2 lost up to 41 mg per drone.

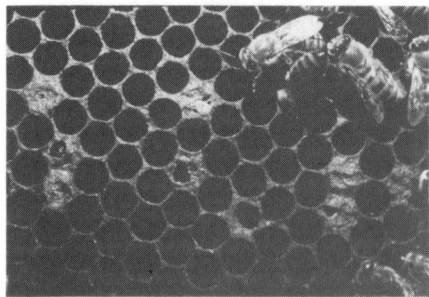

FIG 20 *Worker brood dying during the last stages before the total collapse gives the impression of American foul brood. Varroa mites emerged when the cells were opened. Varroa mites can stay alive for 14 days.*

Yet the most sinister effect of parasitism in drones is the reduction in the size of testicles and their associated glands, as well as in the number of spermatozoa in the semen. The work found that the number of spermatozoa in D1 drones was reduced by 40%, and by 50% in the heavily parasitised drones of the Group D2. Although looking 'undamaged', at least externally, D1 and D2 drones were rapidly lost from the colonies under observation. Many of these drones began to leave the hive at odd times, often during the night. After 7 days (still sexually immature!) only 20% of the drones of D1 Group were left in the hive, and only 10% of Group D2. Drones from both groups were seen leaving the hive and, when attempting to fly, they often only managed to hop. Damage to flight muscle tissue through the mite's salivary enzymes is thought to have been the cause of the inability to fly.

Damage to Queens

Queen cells also can occasionally be

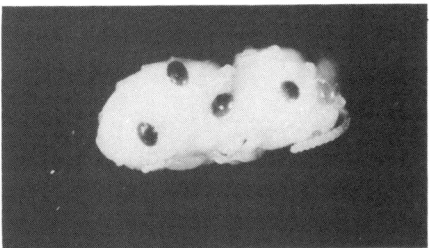

FIG 21 Pink-eyed drone pupa with four mature mites.

invaded when, by early summer and especially around swarming time, the varroa population is too large for all female mites to find 'single accommodation' in a shrinking brood nest. In commercial queen rearing the starter colonies often have very little open brood. They are then given large numbers of grafted queen cells to accept and/or finish. Mature varroa mites then have little choice and will enter queen cells under these conditions. Of course, due to the short period of development and pupation, no varroa nymph will mature in a queen cell, but the damage done by the adult mite and its immature brood is sufficient to render the queen larva a runt and incapable of ever performing as a queen. It is a pity that the work cited above did not examine the number of ovarioles of damaged queens; just as in drones, the hidden damage is affecting the queen's quality, not its looks. Infestation by several mites killed queens in sealed cells and up to 50% of cells did not emerge at all. It is a comforting thought that varroa mites rarely enter queen cells under natural circumstances.

Damage to the Colony

Taking the adult bee next, we must realise, that its hard body shell does not permit the mite to suck blood. The mite must squeeze between the overlapping segments on the ventral side of the abdomen in order to reach, and finally cut through, the thin membrane. Then the mite can draw its 'fill', although each meal is 'only' a small quantity at a time (0.1 mg; a bee weighs roughly 100 mg). No precise reports can be cited as to how often mites must tap the life-blood of adult host bees during autumn and winter, but we can assume that one feed can last several days. After all, no living varroa mite can stay alive for more than 4 - 7 days on unoccupied equipment. Feeding frequency outside brood cells will therefore probably vary between 1-2 times a week. Mites in brood cells with dead pupae can stay alive for up to 3 weeks.

A feed of 0.1 mg may seem a small amount to us, but cumulatively the loss of blood could have a severe effect when a damaged bee's metabolism is further reduced during winter when other stresses aggravate the situation. The aspect of loss of protein from fully developed adult bees and its effect on longevity has not been studied in depth and results could differ from bee to bee and from race to race for the simple reason that bodily reserves of individuals show considerable variation.

Better known is the overall impact of the population of parasites on the whole of the cluster through irritation and disease during winter months (Muller, 1987). Wintering colonies with varroa mites extended brood rearing until late autumn and began to rear brood sooner than other colonies which had been treated with Folbex-VA. Wintering losses of colonies increase out of proportion when varroa is present in the winter cluster. Nosema develops readily in stressed stocks from mid-winter onwards, and many a winter loss is caused prematurely by this disease, rather than the level of parasitation itself.

The problems of the transfer of pathogens by the parasite from bee to bee has been easier to investigate and replicate in the laboratory. After the bee's soft skin between abdominal segments had been pierced by the mite, some blood is drawn from it and and a wound is left behind. Parasitic mites are notorious for forcing salival, protein-dissolving secretions into the wound. The saliva itself may effect the direct transfer of disease. Another action of salival secretions in mites is to keep the wounds open, and these remain a port of entry for invasion by bacteria and virus particles through contamination for some time. Indeed, new information has confirmed, that when varroa mites were transferred onto bee brood, 90% of the killed larvae had acute paralysis particles present in the haemolymph (Koch, W. 1988) He could find no evidence of EFB in the dead larvae.

First evidence of the mite's effect on the overall health of a colony was the finding that particles of acute paralysis virus (APV) and chronic paralysis virus (CPV) could be associated with varroasis and the breakdown of colonies (Ball, 1984). The scientist assumed that natural resistance to the disease is lowered by parasitism. The possibility of a direct transfer of viral particles from bee to bee has not been reported.

Yet any doubt as to the transfer mechanism of bacteria was removed by another experiment which was conducted along the classical lines of bacteriology (Strick/Madel, 1986). In the course of that work *Haffnia alvei bacilli* were transmitted from artificially infected bee brood to other, healthy brood. This was done by transferring initially 'healthy' varroa mites from an infected cell to another cell with a healthy larva. All freshly infected larvae quickly died of toxaemia. It should be noted that *H.alvei* is not frequently found in bees, and so the transfer of this bug from victim to victim (and its isolation from salival glands of varroa!) is firm proof of the transfer of disease from bee to bee.

All findings that Nosema is a fellow traveller in colonies heavily infested with varroa mites must be ignored here. The microsporidian bug Nosema apis (Zander) lives only in the digestive tract and the disease cannot be transmitted via the circulatory system. The above mentioned, intensified development of Nosema is a secondary effect only and possibly due to metabolic and other stresses (constant irritation, loss of blood) experienced by the bees of a colony. When too many bees have been weakened physiologically, when the bee population is too small to form a stable winter cluster, and when this cluster shrinks rapidly from constant loss of bees, Nosema disease is bound to develop explosively. It has been shown that colonies damaged by varroa mites were more likely to rear brood throughout the winter period; Nosema disease may then be the result of the untimely rearing of brood in the winter cluster rather than be caused by varroasis.

Summing up we have to face up to the fact that the effect of loss of proteins during the brood stages of bees results in
1) shortened lives for worker bees
2) reduced size of brood food glands
3) reduced wax production
4) reduced flying activity
5) reduced resistance to common diseases
6) reduced winter survival.

In drones we must add the following
7) reduced number of spermatozoa
8) inability to fly, and, most probably,
9) inability to mate successfully

When we add the direct transfer of pathogens among adult bees of the colony, it can be seen that varroasis can ultimately lead to the total collapse of a colony. When varroa has been allowed to build up over several years into a strong population - even in a powerful stock - the collapse can come about even if no disease is present. This collapse is caused by the sheer number of parasites, and it occurs at a time when no beekeeper would expect it: at the height of the season, July to August.

The collapse is sudden, with bees dwindling at an alarming rate; within three-four weeks there may not be a bee left in the hive. Varroasis is possibly no 'disease'; but it is a killer of colonies when the parasitic populations have outgrown the colonies' natural defence mechanisms and their powers of rejuvenation. Yet all efforts of rejuvenation by brood rearing leads only to an increase in the number of parasites. It is a vicious circle, it is a chain event with only one outcome: the death of the best colonies. So it will pay us to investigate the population dynamics of the mite in the next chapter.

1.5 DYNAMICS OF VARROA POPULATIONS

Distribution, Invasion Pressure and Dispersion

The worldwide and nationwide dispersal of varroa need not be discussed here in depth. The evidence for the efficiency this form of distribution by human action throughout the world is too obvious. Package bees, nuclei with frames of brood, wild swarms in trucks and ships, have all been involved. Migratory beekeeping especially has to carry a fair share of the blame. We suddenly discovered that beekeepers were far more mobile than had been thought.

Postal queen cages must also be suspect, however often this is denied by the parties who are interested in selling queens on an international scale. Huttinger et al (1981) have shown that most varroa mites on bees in cages survived for 10 days, the bees often dying before any mites did so. Sometimes the few, last bees which had remained alive in a postal cage carried 3-7 mites. All varroa mites had died only in those cages in which all bees had died.

Man was involved in transferring varroa from continent to continent and from country to country. The carrier of the parasite over shorter distances was the bee itself. The rapid spread of varroa infestation among the colonies within a district was a new experience for beekeepers. Drifting of young worker bees to neighbouring hives within an apiary had been common knowledge among beekeepers for some time, but the distances involved were usually small. Even older bees were known to become confused when returning heavily laden from foraging trips. Drifting of bees happens all the time, especially when many hives are

a) standing too close together,
b) arranged in one long line at regular spacing,
c) painted the same colour and
d) point in the same direction of the compass.

Young bees on orientation flights and returning foragers tend to drift with the prevailing wind. Even when the winds are not strong enough to cause drifting, tired, heavily laden bees will often enter other hives nearest to the crop. Full of nectar or laden with fresh pollen, these bees lack the 'strange' scent which marks them as foreigners, and they are made welcome in most hives.

When varroa entered the picture, beekeepers of a district suddenly found that their colonies, apparently isolated, were quickly infected by the new parasite. The aspect of drifting was investigated more closely after that experience, and it was found that even bees were more mobile than had been thought possible. Bees frequently entered strange and distant colonies. Even

Bernhard Mobus

colonies far beyond the 'neighbourly' range were invaded by infected bees who had either got lost or were looking for a chance to get past guard bees and steal food. Some races of bees are especially given to robbing because of a well-developed sense of acquisitiveness, and they will distribute mites more rapidly and over far greater distances than other strains. Although research has discovered that very few mites can be found on the older foragers (5 - 7%), no colony in one apiary, after thorough treatment, remained clear of mites for very long, and no apiary remained uninfected in a district once varroa had arrived.

Drones are special guests in any hive and are rarely refused entry in a prosperous colony. Drones fly for miles in their search for congregation areas - and a chance to mate. Marked drones have been found in hives 8 miles from their home apiary. Drones will also carry more varroa mites than foragers over 3 weeks of age. Some work indicates that 23% of the varroa mites outside cells may be on flying drones, while 17 - 20% of the young bees of nursing age were carriers of varroa mites. These latter bees spent most of their time in the brood nest and never ventured far from the hive. Older foragers carried few mites only. Just the same, old or young, drone or worker, in an infested area any 'lost soul' can therefore be a carrier of the parasitic mite, and can make matters worse for all colonies and all beekeepers.

This form of distribution of the parasite through 'dispersion' and 'invasion', through loss and gain can have a devastating affect on population dynamics of varroa, especially in a colony which occupies a preferred situation in an apiary, such as a corner position near the crop or away from the prevailing wind. In such a case the daily arrival of about 10 young mites, soon becoming mature enough to reproduce themselves in the new home, can make a small population grow rapidly to killing proportion. Of course, while some colonies are the receivers of unwanted parasites, it must be obvious that others act mainly as a source, and there the loss of mites can exceed the gain by invasions. Computerised models of population dynamics often include factors of invasion, and these can prove how dangerous it can be when other beekeepers in the vicinity do not apply efficient methods of control.

Swarming is another pathway for the distribution of varroa mites from place to place. Even before varroa arrived in a district, stray swarms were often viewed with suspicion as 'carriers of disease'. This fear was exaggerated, and usually was a fear of 'catching AFB' (a minor risk only with swarms). Once an area is infested with varroa, the problem of 'catching varroa' from a swarm need not be feared by beekeepers within the locality. It is different when the swarm arrives in a varroa-free country by ship or lorry from far away. The constant redistribution of mites within an infected country through drones, drifting and silent robbing is sufficient to re-infect all hives, however good and effective the control methods which the beekeeper applies. Once a country has discovered varroa mites in a small district, the whole country can be declared infected and restrictions can be lifted. The spread of varroa within Federal Germany from a small centre around Frankfurt to blanket coverage of all Lander-states took just over 6 years.

Czechoslovakia tried hard to keep itself free of the mite and killed many colonies in an attempt to stop the enemy. In the end, after killing thousands of stocks and spending much money and effort, it had to give up the fight and learn to live with varroa. The spread of the mite is a relentless advance within a country, bees and mites do not respect borders, even if reinforced with fences, ploughed strips and mines.

Of course, the importation of queen bees into an isolated country which is still free from varroa should be banned. Countries which regularly export queens and bees have, in the past, issued 'Health Certificates' at a time when the mite had spread far and wide - but had not been discovered. Israel, Italy and America are but three examples of countries which declared themselves free of varroasis and exported queens after varroa had arrived. In each case Britain issued licences for the importation of thousands of queen bees right up the the last year before the discovery of the mite. And in each case the thorough search, initiated after what had been thought a 'first strike', found that varroa had covered much ground, and had been in the country for several years.

Population Dynamics within Honeybee Colonies

In this chapter we must combine the facts of varroa's life cycle and behaviour - which we have already learned - and apply them to the life cycle of a colony of honeybees. In the end any computer buff should be able to construct a model of the growth of varroa populations from the facts gleaned so far. Let us enumerate the important factors once more. Some of them can be built into a computer programme, others must remain variables which could be included as "What if..." situations where we enter different values for different situations.

1) Reproduction can only take place in brood cells.
2) Reproduction is possible in worker cells, although drone cells are preferred at a ratio of about 6 : 1.
3) Reproductive factors per cycle vary: in worker cells the factor varies between 0.6 ands 2.5.
4) Reproduction in drone cells is twice to three times more 'efficient', and up to 5 mites can reach maturity (average 3.7 per cycle). On the other hand, most colonies have fewer drone cells, and weak stocks may not have any.
5) Varroa females can, on average, enter cells with brood, (drone or worker) three to four times.
6) Some mites (between 10 - 20%) will not reproduce at each cycle.
7) Some female mites remain unfertilised and cannot produce female mites.
8) The stronger the stock, the greater the number of drone cells.

9) The earlier the colony begins to rear brood, the greater the number of bee generations - as well as the number of reproductive cycles for varroa (1 varroa cycle in worker cell = 13 days)
(1 varroa cycle in drone cell = 16 days)
10) Colony management. Keeping large brood nests throughout summer without a slowing up or a break due to swarming fever, will rear more varroa mites.
11) Any management practices formerly thought of as enhancing the chances of honey production are also the practices which encourage development of more and more parasites.
12) Adverse environmental conditions which could temporarily inhibit or stop brood rearing (drought, cold, strong winds, lack of flowers, pollen and nectar) have an adverse effect on the growth of varroa populations.
13) Racial differences of brooding patterns of bees affect population dynamics of mites.
14) Dispersion and invasion factors cannot be calculated - except as variables.
15) Winter losses are said to amount to around 90% of the autumn population, especially as the length of winters varies from place to place.

European honeybees, especially those colonies in managed apiaries and modern hives, have no natural defence mechanism to win the fight against this parasite. Only few colonies in small cavities which swarm and cast freely very early in the season could possibly survive many years without human help. The outlook for bees in apiaries managed for honey-production appears hopeless without some form of control. The build-up of mite populations can be slow in difficult climates such as in dry, arid regions or in others with long winters and late springs, but it is very rapid in those parts of the world where bees thrive throughout the year.

1.6 NON-COMPUTERABILIA

The factors and variables of development we have mentioned so far have been for normal conditions for bees and varroa. All the same, in nature things are not rigid, and even varroa is not having things all its own way.

For a start, very few colonies rear brood continuously at a steady rate. Brooding is pulsing in fits and starts. Swarming, dearth, even a sudden surfeit of nectar, can affect the number of brood cells available. An ageing queen or the lack of brooding space interferes with optimum brood production for host and parasite. The smaller the colony, or the colder or more severe the climatic conditions, the more a colony is affected by the constant changes within the hive and outside it. Of course, their influence is buffered to a certain degree by colony life within the hive. Just the same, and giving one example only, the cannibalisation of drone and worker brood in times of difficulty, eliminates many mites and their young.

It has been found that in mountainous regions of West Germany the mite does not develop to killing proportions as quickly as in lowland areas with good bee pasture. This may simply be due to a belated start to brood rearing and an early cessation to brood rearing, it can also be caused by other limiting factors. Where bees rely on one or two heavy flows as their main stay in life, with drought and dearth restricting brood rearing throughout the rest of the year, the influence on varroa populations is quite marked. Just the same, it is impossible to quantify such factors for a computer programme - and it is dangerous practice to rely on vague arguments for prediction or treatment.

Another uncertain factor which can upset a programme is the following observation: Sometimes only female varroa nymphs are born in one cell. No male mite is then available to fertilise the females. Only male mites are produced in the following cycle when the uninseminated mites enter a cell and become gravid after their first meal. Prof. Woyke was convinced that in such a case the first male 'son' is able to fertilise his mother mite. Entering another cell, she then would be able to produce female mites in her second cycle. New investigations made in Holland disprove this, as older, unfertilised mites, though repeatedly in contact with male mites over a number of cycles, carried on laying unfertilised eggs and producing male progeny only.

Some mites are unable to produce any eggs in worker cells at all - or only after a lengthy delay. All nymphs of mites reared after a delay will be immature when the worker bee emerges; only mother varroa will remain alive and try again. Lack of larval hormones may be influential in such a case. After all, it is the lack, or the low levels, of this hormone which prevents reproduction of mites in worker cells in colonies of the Indian bee. Some workers believe that this is the mechanism which holds varroa populations in check in that honeybee and at a level which ensures colony survival.

We have seen (**Chapter 1**) that the duration of the bees' pupation period is another variable factor of influence on the reproductive rate of mites. This must make us investigate the pupation period of worker bees in greater detail. Many bee books tell us that worker cells are sealed on the 8th day, and that the young bee emerges on the 21st day after the egg had been laid. This is a rough and ready guide; yet many bees emerge well before 21 days are up, while in other cases the development is slower and varroa scores another 'double'. Delayed development is a general occurrence when brood nest temperatures drop slightly in the hive towards the end of the year. The same happens when thermal economies must be made during a lean and hungry period. Another low temperature period comes about when a swarm with queen has left, and finally no open brood remains in the parent stock. In one such case the swarm with a marked queen had been taken and hived 27th July When the parent stock was examined at the 29th August, worker bees were still seen leaving cells after 31 days, with more worker yet due to emerge. A virgin was present by then, but no eggs had been laid, and there had been no chance of another queen having laid eggs after the swarm had issued. Under these conditions even worker brood would have provided conditions for a rate of reproduction far in excess of the factor of 1.00. On the other hand, no open brood had become available in that stock at a time when summer mites are short lived and cannot reproduce.

Accelerated development of bees can be a factor with a negative influence on population dynamics of varroa. In one case an emergency swarm (no queen cells, no queen cups with eggs in the hive) left a hive on a hot day. Good weather and copious flows of nectar followed the hiving, and the whole sequence of hatching, of larval and pupal growth was speeded up considerably. Many cells were already sealed by the seventh day after hiving, and emergence of first workers also occured days sooner than normal. The pioneering enthusiasm of a swarm had speeded all metabolic processes.

Apart from delayed or accelerated emergence for the above reasons, minor differences in emergence from pupation exist between races of bees, with most European bees varying around 21 days for full development from egg to adult bee. The Cape bee, *A.m.capensis*, on the other hand, has the shortest period of 19 days. In that case, very few varroa nymphs in worker cells will become mature enough to survive in the outside world. Varroa mites struggle along until another flush of drone-fed mites can bolster up dwindling populations. Varroa cannot become a dangerous parasite in this race of Cape bees - if it can survive in that race at all.

The New Varroa Handbook

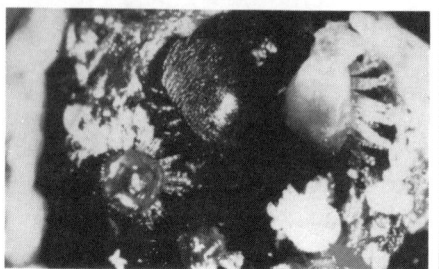

FIG 22 View of bottom of a cell after the emergence of a worker bee. Immature varroa and faecal pellets are clearly visible on the bottom of the cell.

Some strains of bees suffer more from chalk brood than others, and drone brood especially is a vulnerable victim of the disease. When varroa prefers drone brood by a factor of 6 : 1, some mites are bound to remain trapped in cells with chalk brood 'mummies'. The disease could therefore have an effect on varroa populations, but this too must remain an unquantifiable factor; it is too vague for calculations.

Bernhard Mobus

DETECTION OF VARROA 1.7

'Instant' Tests and their Reliability.

Fighting the enemy is not a problem in countries without varroa. On the other hand, all interests of the alert or frightened beekeeper are centred on the early detection of the mite's arrival. Amateur beekeepers and commercial beekeepers in these countries are encouraged to discover the 'first strike', so that 'counter measures', if any are contemplated, can be introduced at once. These would range from the attempted elimination of the 'only' affected colonies in an apiary to the introduction of registration of beekeepers and of regulations which govern the migration within the boundaries. Nowhere have such regulations proved effective in protecting the rest of the beekeeping community. All measures came too late to be effective, and all regulations which were introduced in the initial panic have, in the end, become cumbersome restrictions which had a severe effect on the work and the income of professional honey producers. Furthermore, no country has yet been able to clear up a new invasion.

Very few parts of the world are nowadays so isolated that nothing can slip, unobserved and uninspected, over the borders by air, sea or road. Island countries and continents (Australia, Britain, New Zealand, Ireland) do enjoy limited isolation, but beekeepers everywhere must be prepared for the day when the mite arrives or, better, is detected within the boundaries. Thorough investigations on a large scale, which were introduced after a new find, have always established, that the mite had spread far and wide and had been present for some time, often for years. By then it had become impossible to discover the initial importation with any degree of certainty. Fingers were pointed in all directions, and rumours made matters worse by spreading suspicion.

FIG 23 *A floor board of a 'National' hive with paper inlay and varroa screen. (Note the setting back of the front member of the screen to keep the entrance open at all times).*

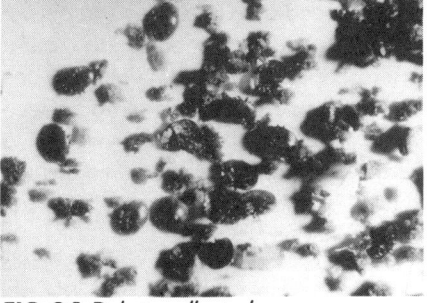

FIG 24 *Debris collected on a paper insert. Shiny, oval varroa mites can be seen among wax particles and other detritus.*

37

The New Varroa Handbook

In order to discover its presence, countries without varroasis encourage their beekeepers to make annual, diagnostic examinations by means of simple 'instant' tests. The use of paper inserts under a screen stretched across a simple frame (2mm mesh, 10 mm above paper) is recommended. The inlay remains on the floor throughout winter and is examined in spring for mites which had dropped off the winter cluster. Without a screen the method of detection of a first strike by a paper inlay is unreliable.

The examination of paper inlays by laboratories of governmental extension services is not a simple task. The inlay holds all the dross, debris and bees which fall from a winter cluster, as well as small insects which have sought refuge

FIG 26 *Installing the paper insert for placing on bottom board.*

in the hive. Damp and condensation fuses the debris into solid cakes which make examination difficult. Honey-sugar granules are easily eliminated by dissolving in water, wax granules require a better solvent and petrol has been used in the past. the dross is then passed through a number of sieves with the last one retaining all varroa-sized particles. Experienced eyes must then sort out the debris, usually under a dissecting microscope. When the value of the method of detection has already been reduced by using an inlay without a screen, the expense of examining winter inlays is nothing but a waste of tax payers' money.

Following experiments made by Dr. de Ruijter in the Netherlands, Britain also recommends an 'instant' test by means of which the chances of detection are

FIG 25 *Screened paper insert on bottom board for the detection of dead varroa mites.*

enhanced and the amount of debris is reduced considerably. Advisory leaflets suggest the use of tobacco smoke as a means of detecting an outbreak of varroasis in the early stages. This method consists of blowing the smoke from 3g of tobacco through the entrance of the hive in the evening. A paper inlay is required to catch the killed or anaesthetised varroa mites. The paper inlay is collected after a few hours or the following day and is sent away for analysis by the governmental extension service.

Because dead or unconscious mites are very light, the fanning action by bees will remove many mites from the paper inlay before it is collected by the beekeeper. Hive cleaning activity by bees will also remove dross, debris and mites and thus reduce any chances of detection. The method is therefore unlikely to discover an infestation in the first year and before the mite has multiplied and spread. The use of a floor screen would improve the method by preventing bees having access to the paper inlay, - and incidentally, also preventing stunned and recovering varroa mites having access to bees. Coating the edges of the paper inlay with Vaseline (soft petroleum jelly) would trap varroa mites which are recovering from the effect of the smoke. But again, without varroa screen and Vaseline the above method is unreliable as a means of finding a first strike of varroasis. It is also a method which is laborious and often involves a change in the design of the standard floor board.

In Britain the number of inlays sent to the official stations has been very low in the first two years. Apathy, or a reluctance to subject colonies to tobacco smoke, won hands down over the fear of varroasis. The high duty on tobacco did not help either. Of course, many chemicals other than tobacco can be used as diagnostic tools. Again, whatever drug is applied, a screen and paper inlay on the floor should be used for catching the anaesthetised or dead mites. Amitraz fumigation, either as a strip or aerosol, Folbex VA, Perizin and Fluvalinate, in fact, all efficient chemicals which have proved their worth in the control of varroasis, belong to the category of proven diagnostics and will increase the chances of early detection. No country has adopted tobacco smoke as a good method of treating colonies infested with varroa, and its use lowers the value of all efforts of detecting a first strike. Again, good diagnostic chemicals, combined with screened inlays must be the solution for early detection in an apiary.

Some countries have recommended a method which involves the killing of a larger number of adult bees. About 500 bees are caught in a glass jar and are given a short burst from an aerosol can with a mixture of ether/alcohol mixture (such as is used to start Diesel engines from cold). The mixture makes bees disgorge the contents of their honey sacs, and varroa mites then adhere to the sticky sides of the glass jar as it is turned by the operator. Here they can be seen easily. A similar method involves the 'washing' of the dead bees in petrol or in water with a few drops of detergent, then shaking and filtering everything through a screen of 2mm mesh size which retains bees but lets varroa drop through. Because even 500 bees are only one percent of a strong colony, the odds are stacked against an early detection of an outbreak, especially so when most mites are hidden within brood cells. This drastic method will not find favour with beekeepers and is inefficient when we look for a 'needle in

FIG 27 *Three mature mites and an immature deuteronymph on a purple-eyed pupa.*

a haystack'. A quick calculation should tell us that if we find one varroa on 500 bees, then it is unlikely to be the only one in the hive. As many as 99 more could be on the rest of the adult bees and a further 9000 mites are probably hidden in the cells of brood. Such a case is clearly no case of a 'first strike'; that colony would be doomed to die in the following winter. Crippled worker bees would have been seen crawling from that hive long ago.

A reliable but laborious method of detection is the examination of sealed drone brood when the pupa has reached the dark-eye stage. By then many varroa nymphs would be fully coloured and can be seen with the naked eye. As a strategy for early detection it is far too cumbersome - and too disturbing for the colony and the beekeeper. Uncapping sealed drone brood is of greater value for establishing the population density of the mite, but most beekeepers would shrink from that task. Once a colony is infested, the removal of drone cells will probably have its followers in the hope of a cure or of a slowing down of the rate of increase in numbers. Sealed worker brood is even more difficult to examine cell by cell, and is unlikely to find favour with any beekeeper. The method can serve its purpose as an investigative tool for the scientist who is studying population dynamics or is looking for genetic variations of resistance to varroa.

Routine Assessment of Varroa Populations

On the whole, the paper inlay, used in conjunction with a screen and some diagnostic treatment, is the best tool for monitoring current varroa populations. Once the mite has become established in a colony, it is important to make routine assessments of the growth in the number of varroa mites. Without such knowledge the beekeeper could be wasting time and money by treating colonies with a low infestation, or he may neglect the urgent application of remedies when the parasitic population is approaching killer proportions. The question of applying expensive treatment annually or biannually, singly or on a blanket basis will remain a problem for professional beekeepers as long as the available chemicals are costly. Routine surveys can be made without the use of chemicals, and the weekly death rate of mites can be assessed as long as a screen and inlay is constantly in place - maybe under one hive in ten. It will catch and retain any mite which dies a natural death and drops off a bee or out of a cell after the emergence of a young bee.

A new design for a hive floor, developed for quick and easy sampling, would ultimately be the answer to permit constant surveillance of mite populations. The design should make withdrawal of screen and inlay - possibly on a shallow tray of plywood or plastic, an easy task. Counting of mites - not very difficult with reasonable eyesight, allows the beekeeper to either leave well alone - or take emergency steps of control. That can save time, money, and colonies. The tray can also be used for fumigation with formic acid or other varroacides from below. Such a floor should be deeper than normal, with the surplus space possibly separated from frames by means of a slatted floor. The space under the slatted floor can then take a tray with screen and the paper inlay. Withdrawal from behind would be an improvement, but is not strictly necessary. It would become popular once beekeepers realise its value for improved wintering by providing a larger air space under the frames.

A weekly inspection of the floor during summer would let the beekeeper count the dead mites and give him an idea of the total population and its rate of development. He can then decide to apply suitable treatment at once in order to make sure a relatively sound winter population of bees will form the winter cluster, or risk leaving treatment for another time. Blanket medication can be avoided or left until all danger of contamination of honey is over. In the main, the monitoring of varroa populations by any of the above methods will remain restricted to the small scale beekeeper.

Research on the interpretation of sampling without the use of diagnostic chemicals has to be done for each country or, better, for each region. Population dynamics of varroa varies to such a large extent from hive to hive and district to district, that it is impossible to forecast the actual number of mites in each hive when more than, say 50, dead mites are found on the inlay per week. Work done in Germany, with the Carniolan bee as the dominant race, will not necessarily be applicable in countries where the Italian bee, or selected, named strains of that race, are preferred by commercial beekeepers. Some work suggests that colonies are nearing break-down point when more than 10 mites per day are found on the inlay during summer months.

As things stand, the commercial beekeepers will be faced with much additional investment and labour costs when varroa has invaded a country. They cannot make routine inspections of thousands of colonies. The amount of labour can be reduced if only a small number of colonies, picked at random in each apiary are equipped with a varroa-sampling floor, and the whole apiary is then blanket-treated according to the needs of the few. Yet large scale beekeepers will even find such work an unacceptable, costly task. They will probably plummet for blanket treatment as soon as it is known that varroa is within the district, especially when a cheap version of a reliable varroacide becomes available.

1.8 THE CONTROL OF VARROASIS

Tackling this chapter in order to supply definitive information will prove a most difficult, nearly hopeless task. It will probably never be a closed file. The picture is constantly changing, and every day brings news and hopes from another part of the world, only to be dashed again in a later report. The trouble is that varroasis is a 'new' disease. The known bee diseases have been with us for a long time - some of them were known to the Romans. Even the honeybee had 'learned' to live with these 'old-fashioned' diseases, and Nature constantly selected the existing bee populations for survival mechanisms which were geared to overcome them. Even the dreaded AFB was no problem to bees in the wild, and bees shrugged off the problem through hygenic behaviour of cell cleaning activity in the hive, through starting afresh after swarming and casting - and through death when too weak. The mouse and moth brigade then cleaned up the remains.

The European bee has now become a host to a parasite it had never encountered before. It is defenceless, it has no behaviour patterns to deal with the parasite. Although we must remember that without human interference the problem would never have arisen in the first place, vast stretches of the world would become devoid of honeybees - and varroa parasites - without human involvement now. It is even possible that 'European' bees would have to re-evolve from mite-resistant southern races which, in turn, would have to conquer Europe a second time by learning to survive harsh winters as tight clusters. 'Wintering' itself might have to evolve anew. Like the monkeys in a cartoon, overlooking a landscape davastated by an atomic holocaust, *Apis mellifera*, in all its various races, would have to start all over again after the varroa holocaust.

Because the honeybee is so important to the environment and to human agricultural interests, the countryside would suffer greatly without the bee. Beekeeping is big business when seen as a whole. It is also little business, it is a way of life as well as a hobby, and so every beekeeper will make an effort to keep the bees alive and maintain his or her income, however large or small. At this moment we have to admit that this will not be an easy task, and it will involve constant work. Just the same, there may be a light at the end of the tunnel, however long it seems, and I do hope it is a bright one for all beekeepers.

Basically all measures of control can be categorised as follows:
1. Biological or Bio-technical Methods
2. Physical and Mechanical Methods,
3. Drugs and Chemicals

Each category can be split again into separate fields, and we will do that as we go along.

Bernhard Mobus

1. BIOLOGICAL METHODS OF CONTROL

Genetic Selection of the Honeybee

This chapter will deal with 'cures' which do not make use of chemicals or applied physics. They may involve manipulations or management techniques which are based on the knowledge of the enemy's biology.

The most promising, the most safe, and most effective long term control of varroasis would seem to lie in the field of genetics, or when a chance mutation occurs in our European bee. In the Indian bee, *Apis cerana*, varroa mites are unable to reproduce in worker brood. Female mites may enter worker cells, but they remain trapped for the full period of the bees' development. The mites are unable to even lay an egg. This fact alone reduces the mite's chances considerably and new generations can only be produced when drone brood is present in the hive. Until then the mites grow older and older and fewer and fewer, and populations remain at the level which can be tolerated under the circumstances which prevail in India.

Selecting bees for the factor which inhibits reproduction in worker cells would be an ideal solution, but so far it has not been found in any colony of European bees as a dominant genetic character. In Uruguay a beekeeper has found that his bees 'can live with varroa' without medication. In Yugoslavia a similar event happened, but the progeny of the colonies did not show any 'resistance'. Maybe the dream will be achieved one day when a colony survives in spite of neglect - and all later generations possess the same character in a dominant gene.

Another defence mechanism against the mite has recently been discovered in the Indian bee. The worker bees will tackle varroa mites, chew them between mandibles and carry them out of the hive. This behaviour mechanism is absent in the European honeybee and is unlikely to develop in time to do us any good. After all, bees have been unable to get rid of their bee-louse passengers. When dreaming of including that gene in our bees, we should be aware that the Indian bee is another species, and that crossings between the Indian bee and its European cousin (twice removed), - even using instrumental insemination, - have all been unsuccessful in the past.

The Cape-bee (*Apis mellifera capensis*) is a race of the European bee, although some workers tend to classify it a sub-species. It lives in a narrow coastal belt along the Cape of Good Hope of South Africa. The period of pupation is a very short one, and the bee can complete its full development from egg to adult in about 19 - 19.5 days. No fully-coloured, mature varroa mites were thought to be able to emerge with the mother mite when the young bee emerges. Again, it is the inability of varroa to succeed in reproducing itself in worker cells which is supposed to make that race of bees 'immune'. Hopes to incorporate this African bee into a breeding program for a 'resistant' bee were based on old observations. However, newer work (see

"Varroa Up-date"), shows that the mites can reproduce successfully in a colony with a 19-day development, albeit with reduced success).

Let no one get the idea that the importations of Indian bees or Cape-bees would be the salvation of beekeeping. Although cross-breeding is feasible in the case of the Cape-bee, but we all know where the introduction of another race of African honey bees into Brazil led us. In any case, the Cape-bee and other African bees lack the ability to winter, and no attempt should be made to play around again with races of uncertain positive and other, definitely negative genetic values.

From now on every beekeeper must learn to 'see' where he was only 'looking' before. He must learn to 'read' the colony's progress and behaviour. He must assess the populations of varroa mites within the hives and must act according to regulations - and his powers of observation. In many districts there are colonies with behaviour patterns or characteristics which show variations from the norm - within the genetic range, of course. Selectively breeding among the 'fast breeders' can possibly lead to a shortening of the sealed brood stage and, hopefully, to salvation or a diminished threat. What happened in Uruguay (beware, it may have been due to climatic conditions) and in Yugoslavia can happen here. We too may come across a colony which can cope with varroa. It is not certain which trait, or combination of characters will lead to that success.

Any beekeeper who comes across a colony which is able to 'tolerate' varroa without 'help', should donate that stock to a research station to be set up by Governments for further study and for experimental breeding. Although he should do so from altruistic motivation, he should be highly rewarded for such a good deed. We must be aware though, that some wild colonies in small cavities need have no defence mechanism for staying alive; their survival is entirely due to the difficulties which varroa faces in a colony with a small brood nest, with a slow start in spring and which swarms and casts naturally at an early stage of its development. All the above factors are introducing long brood stops at vital times during which the mite cannot reproduce itself. Instead, varroa ages and dies all the time. But it is an obvious statement that keeping weak colonies is not conducive to honey production.

Dr. Maul speaking at the Freiburg Conference (see **"Varroa Update"**), forecast that selection for some form of resistance is not such a far cry. Such work had been fraught with difficulties, yet re-queening with queens from a colony with low rates of reproduction of the parasite gave encouraging results in some colonies. Just the same, let no one in Britain dream of the day when he can live with varroa "nae bother"! A chance mutation can occur tomorrow, but Nature usually takes its time, and measures time in terms of millions of years of evolution.

Selecting for genetic 'attractiveness' of worker or drone brood is another approach. In this connection we could have two antagonistic characters, and finding the right one would make the choice difficult. Selectively breeding for worker brood without any attraction for varroa would make more mites wait for drone brood. A bait comb of drone brood would then catch and eliminate most mites. It is a pity that such a strategy seems a hopeless task. At the moment, by constantly removing varroa mites which prefer drone brood, we are actually selecting a strain of varroa which prefers worker brood!

Now the other side of the coin. Bretschkow did notice in one apiary of Carniolan bees in Austria a higher incidence of infestation of worker cells when compared with the usual preference for drone cells. He found that the varroa population stabilised itself at a level which proved 'non-lethal' for the colonies. Too many varroa mites were chosing to reproduce in worker cells, and the population explosion which only comes from the greater rate of reproduction in drone brood, did not occur. That observation needs following up to see if local, environmental conditions are the key to his 'success'.

Bio-Technological Manipulations of Colonies

Hoping for the bee or varroa to change its ways is one approach, exploiting varroa's biological needs, urges and failings, - and then killing it, is another. This approach involves contrived manipulations, a lot of work, as well as good record keeping. Catching many mites in combs of worker or drone brood and destroying (or treating) the parasite while it is held captive, has produced good results. Other methods of bio-technological control rely on interfering with the bee's patterns of brood rearing to the detriment of varroa's population dynamics within the colony.

The Brood-Stop in Mid-Winter

One approach on this line was the idea of caging a queen in a winter cluster. This would prevent untimely brood rearing, and only the release of the queen in spring would give varroa a renewed chance of reproduction. By extending the broodless period of winter, the varroa population was expected to diminish sufficiently to represent no threat any longer (Alber, private communication). Scottish investigations (Mobus, 1979) have shown that the caging of queens in a winter cluster is not without risk to the health of the colony. When queens were caged in mid winter in a cluster with brood, the colony developed dysentery within a short time. The rearing of brood in mid-winter is the colony's way out of a dilemma when a 'positive' water balance of a hive-locked winter cluster becomes an embarrassment to the bees. The elaboration of brood food, combined with higher cluster temperatures, can bring about some relief of the water problem. Eliminating all chances of brood rearing by enforcing a brood stop in mid-winter could aggravate a hidden water imbalance until dysentery is inevitable. Applying the bio-technical cure in mid-winter makes matters much worse.

The New Varroa Handbook

Preference for Drone Brood

One of the first hopes for successfully slowing down the rapid development of mites was the realisation that the mite prefers drone brood to worker brood when chosing a 'home for its babies'. Cutting out patches of drone brood, once fully sealed, was tried initially with great enthusiasm. It is still done now, but early hopes for an efective control did not materialise. Adding frames with drone foundation or starter strips was a modification of the scheme, and these drone bait combs were removed and destroyed when fully sealed. Of course, the beekeeper hoped that most varroa mites had voluntarily gone into the trap. Successes were reported time and again but, the method was found to be only partially successful in a number of carefully conducted experiments (Ruttner). When mite populations were very low, this trick kept the number within acceptable limits. When the infestation had risen, the removal of drone brood did not fulfil expectations. However, this line of control is still followed by many beekeepers who abhor chemotherapy.

Varroa Bait Comb

When it was realised, that varroa females can also reproduce satisfactorily in worker brood, the use of the 'Varroa Bait Comb' (worker pattern) was investigated. In order to eliminate competitive brood on other frames, the queen was confined to an empty, drawn comb within a frame cage made of excluder material. After 7 days the first frame was removed from the cage and another drawn frame was substituted and, again, the queen remained caged on this frame for another 7 days. The bait combs (marked) would be full of sealed brood 18 - 19 days after the start of the operation and, hopefully, full of varroa mites trapped in the brood. Instead of using a frame cage in order to confine the queen, the same goal can be achieved by one or two tightly-fitting divider boards with large areas of excluder material. The queen is then confined to a compartment within the hive. When all brood is sealed, but before any has emerged, the combs are removed and destroyed. The queen is released again to carry on as normal.

In the early days this proved quite successful and removed many mites, together with their brood, from the colony. It reduced the pace of the mite's development and kept the level of infestation within 'acceptable' limits. Just the same, the removal and destruction of so much worker brood is not a happy solution for beekeepers. Some of them built large incubator cabinets for holding many infected frames. When brood had emerged, the young bees were then treated with an acaricide before being used as a shaken swarm, or they were used as package bees with a young queen. Since those early days, the effect of formic acid on sealed brood has become known as an alternative to destruction, and the varroa bait frames can be returned to the colony after 90 minutes in a special, insulated fumigating box.

Timing of the operation can be adjusted to suit colony conditions or management programs. Indeed, it can be introduced as routine work of swarm control, and the bait combs (with adhering bees) can be used to form nuclei, later to be given a queen cell. Fumigation can be applied before the virgin queen has sealed brood of her own. It is worthwhile to remember that the young bees were damaged by the blood-letting, and need special care and attention.

Artificial Swarm

Ritter reports a scheme which, without resorting to the use of chemicals during the summer period, uses the bait comb method of eliminating varroa from a colony and overcoming swarming tendencies at the same time. He suggests using the "Artificial Swarm" method of swarm control. This is done by setting the colony aside on neighbouring stand, bringing a second hive with foundation or drawn comb (no brood) to the original location, and returning the queen (only!) to this new brood chamber. Foragers and older nurse bees return to make up the artificial 'swarm'. Giving this colony a varroa bait comb of eggs and young larvae, Dr. Ritter hoped to remove many mites. He was surprised when only very few mites were counted in the bait comb. So few in fact, that he now no longer recommends giving the artificial swarm a bait comb. he found that flying foragers are host to very few varroa mites only and that most of the parasites had remained on the young bees, the nurse bees and in the brood cells of the queenless 'parent stock'. So he changed his advice and proceeds as follows.

Make the artificial swarm as described above. After 9 days (no open brood left), the parent colony must be inspected and all queen cells EXCEPT ONE are cut out. This one cell is retained in a nursery cage to which bees have access; this avoids the colony getting frustrated. After 3 weeks (24 days if drone brood is present) all brood in this stock will have emerged together with all varroa mites. Two bait combs with eggs and open brood are now transferred from the queenright 'artificial swarm' to the 'parent stock' with the young bees and the mites. The virgin, now emerged, remains caged. When bee brood is fully sealed - and before any bee emerges, the bait combs are removed and destroyed (if no fumigation is intended). A new, mated queen is now introduced to this stock after the removal of the caged virgin. If the work is done at the right time, this method successfully suppresses swarming.

Variations on the theme can be devised, such as adding frames of foundation routinely in late spring and getting the queen to concentrate all her egg laying on these few frames. After 18 days the frames are removed, together with adhering bees, to form a nucleus, and a queen cell is added. Of course, the introduction of mated queens (instead of a cell) is more attractive under normal circumstances, but this perpetuates the situation without relief and can only be done when chemotherapy is used. Otherwise the mites can enter more fresh brood at once. When a queen cell is used, there will be a period without brood during which varroa mites will age and die to some extent.

The New Varroa Handbook

Introducing a Brood Stop

Indeed, the very simple method of re-queening by de-queening and letting the colony rear a new one - if possible after a delay - has proved a good plan in some countries. It is exploiting the natural death rate of varroa. This method of control is usually done when queens are old. Young queens, on the other hand, may be caged in such a way (Worth - cage) that bees can lick and touch her, but she cannot lay eggs for a time, usually two to three weeks. The life span of the summer generation of varroa mites is short, and many of them will die during the break in brood rearing. For varroa, a gap of two weeks amounts to a loss of one full reproductive cycle. Three weeks add a further half-cycle and this will ensure that nearly all sealed brood has emerged, thus exposing all varroa mites to potential medication at the same time.

A 'rested' colony, with little open brood over such a period, will raise its first frames of brood with renewed vigour. Temperatures of the small, but rapidly growing brood nest are easily kept at an optimum level, and the provisioning of new larvae will be intense. The whole development of this first generation of bees is speeded up, and varroa may well be less successful in its first attempt to reproduce successfully. It is not difficult to combine a brood stop with a varroa-baiting scheme, catching again many 'randy' varroa with the urge to reproduce, and controlling swarming effectively at the same time.

Such a planned brood stop is of great value where long-lived bees are maintained. Where highly prolific, therefore short-lived bees are the dominant race, colony strength will suffer badly and a change-over to another race should be contemplated if this approach to controlling varroa, of keeping it down at relatively harmless levels, is to be successful. No promises of success can be given here; what succeeds in one district may easily turn out to be a failure in another part of the world. Ultimately, unless a cheap, quick and safe chemical becomes available, it must be some form of biological control which will make sure that bees survive and honey remains entirely wholesome.

2. PHYSICAL AND MECHANICAL METHODS

Heat

The chapter is a short one. Until recently only the use of higher temperatures, above those of the brood nest, have shown promise. Controlled heat may be applied to bees or brood, but the work is fraught with difficulties.

When varroa mites were experimentally exposed to a temperature gradient, the mites preferred normal brood nest temperatures of 32 - 35°C (90-95°F). Given a free choice, the mites rarely entered the zones of temperatures above 38°C (100.5°F), and they died when temperatures rose above 42°C for a while. Although heat treatment to kill varroa appears to be a 'simple' solution, a colony with brood will efficiently counteract any attempts to raise temperatures by fanning and the evaporation of water. Any experiment to control varroa by heat in a colony with brood must therefore fail.

Adult bees without brood, as a natural or shaken swarm, have been subjected to temperatures of up to 48°C for varying lengths of time and the treatment decimated or eliminated the parasitic mites. Bees do not like such heat treatment and begin to panic in enclosed situations. They tend to block screens of cages and make matters worse. Large clusters can suffocate if the area of ventilation is insufficient, and any heat treatment is best applied by means of a forced, thermostatically controlled air stream to bees in a screen-cage of adequate dimension. A small amount of scented oil, Thymol or similar, triggers cleaning behaviour in bees and helps to break up the cluster for faster and even penetration of the heat.

Sealed frames of brood without adhering bees have also been exposed to higher temperatures in an effort to kill varroa mites within cells. All nymphal stages of varroa as well as 80% of all mother mites died in brood cells which were exposed for 12 hrs to 40°C (104°F), or for 4-5 hrs at 44-45°C (113°F). Early stages of sealed bee brood was slightly more sensitive to the raised temperatures, but the overall losses of bee brood did not exceed the 5% limit.

All heat treatment requires skill, a lot of labour, extreme care and costly apparatus. On the other hand, it is a method of control without any problems of residues and damage to honey. Bees may well be physiologically damaged by exposure to these temperatures, but no investigations into that aspect have been reported.

Dusts

A promising development has been reported recently. Investigations (Ramirez, W.B. 1988) have shown that dusts and powders can be applied with good results. His study was suggested by the common knowledge that birds enjoy dust baths. All feathers are thoroughly filled with fine particles - and irritating parasitic mites are lost when the bird shakes its ruffled feathers.

The action of dusts, toxic or non-poisonous, on varroa mites can easily be explained. Varroa mites have no claws, but have sticky pads to retain a foothold on bees. When fine dusts settle on the surfaces of even one pair of pads, varroa is said to be unable to cling to its host. The mites slide off, some of them bouncing into open cells, others falling to the floor. Here they should be removed before they can regain a foothold. Better still, the floor can be removed entirely before treatment or it can be replaced with a framed screen. French workers also have examined this approach to the problem of varroa control, and 100% success was achieved 14 hrs after applying finely powdered glucose to shaken bees. One brand of powdered pollen substitute had similar success rate, and even pollen dust achieved 87% control. Talcum powder, finely powdered leaves and pollen substitutes were all nearly equally effective, but must contaminate honey stores to some extent.

Very few objections can be raised provided a suitable dust is used. Applied by hand, it will be a laborious process and should be done during a season without brood in the hive.

FIG 28 *Blowing chemical fumes from a smoker into the hive entrance. This lends itself to the treatment of several colonies with Folbex VA, phenothiazine or similar fumigants.*

The basic method permits mechanisation and, in spring, powdered pollen substitutes or powdered glucose, maybe even icing sugar, could be applied to all colonies with the aid of a bee-blower. Little harm will be done to bees and honey. For large outfits the news is possibly a bright light at the end of a tunnel when chemical treatments are abhorred or are illegal. Just the same, the method needs further investigation about the maximum size of the dust particles which make powders effective.

The recommended summer treatment (with brood) is for six applications of 50 g powder given at at intervals of 4 days (max. 7 days), thus catching most of the young mites during the maturation period and before they enter brood cells. When foul brood is present, the neutral carrier (even icing sugar is suitable) may be mixed with Terramycin, - provided no honey flow is in progress or is expected shortly. Natural or shaken swarms, or bees without brood, need fewer applications.

3. CHEMICAL CONTROL

It is strange that the beekeepers who are the first to complain of, and are most vociferous against the use of chemicals in the environment, suddenly look for a a 'Magic Bullet' in their fight against varroasis. Beekeepers will have to realise that they may have to rely on chemicals for the salvation of their colonies, because few of the above mentioned manipulations and treatments are easily applied on a large, commercial scale. The most 'natural' approach, that of the varroa bait comb combined with a brood stop, will prove unattractive to all beekeepers whose ideal colony is the one which has wall-to-wall-carpeting with brood and bees throughout summer. Yet the chapter "Population Dynamics of the Parasite" shows that the most prolific strains of bees suffer most when varroa enters their lives. Such colonies will be fighting a back-to-the-wall battle. No wonder beekeepers all over the world are looking for a 'Magic Bullet' - especially one which will kill varroa, will harm neither bee nor anything else - and will leave honey uncontaminated.

When the first heavy colony losses were experienced in Russia, experimenters reached first for the material they had used before and which was known to be relatively harmless to bees. Fumigation

Bernhard Mobus

with phenothiazine had been recommended in Russia for years against the bee louse *Braula coeca*. The material was also relatively cheap, freely available, easy to use and achieved 'kills'. Of course, counting dead mites under the frames was all that mattered at first, and every mite found dead was hailed as another success.

Only when Germany discovered varroa in a few colonies, did any one bother to count the number of varroa mites which had survived a treatment - however many had been killed. Only by counting the survivors and comparing their numbers with the number of killed mites can the true effectiveness of a chemical be assessed. Of course, counting varroa 'survivors' required the killing of whole colonies and the 'washing' out of parasites left behind after treatment. Varroa having no claws, but sticky pads, the washing proved effective and accounted for more than 95% of the survivors which had been on adult bees. But such an investigation demands the sacifice of many colonies. Furthermore, the careful counting of all varroa mites left alive in brood cells was also a necessary and laborious process, but this scientific approach proved how futile any chemical treatment during summer months could be.

After the discovery of varroa in Germany, Dr. Wolfgang Ritter was given the task of testing many chemicals for their effectiveness against the parasite as well as its bee-compatibility. Application of chemicals, at various concentrations, was done by fumes, sprays, dusts and, later, by feeding. Some of the compounds tested had to be used in several ways and gave different rates of success. Not all experiments can be mentioned here, and the picture is changing constantly.

Gases

Carbon dioxide was tried in Greece in the beginning but achieved little success. Bees and varroa became anaesthetised and fell to the bottom of the hive. Many bees were killed when lying in a heap on the floor. Varroa in brood was not affected, but bees had aged physiologically and could not rear brood very well after treatment.

Observations of the behaviour of birds in the wild came to our aid again. It has been known that they like to take a 'dust-bath' in ant-hills in order to get rid of mites. Other birds had been observed to pick a beakful of ants and to rub them under their wings. The venom was a natural insecticide and acaricide and killed lice and mites among their feathers. This suggested the use of formic acid for extensive trials.

Formic acid is normally a liquid with a very pungent smell. It evaporates easily above 10°C (50°F) and becomes a gas. The vapour molecule is very small and can even penetrate through the minute pores in cappings which let the brood breath. Indeed, formic acid, as a molecule, is only slightly 'heavier' than CO^2, and where one gas can escape from the cell, the other can enter just as easily. Formic acid dissolves in water, and open and sealed honey stores absorb some of it readily during treatment. Yet formic acid evaporates

51

The New Varroa Handbook

again from the honey, and after a few months the levels of acid approach those found normally as a natural constituent in honeys and honeydews (up to 1 g/kg).

The countrylore suggested trials, and small bottles with wicks were tried first. In old-fashioned German hives it was difficult to maintain even concentrations of the vapour, and results were varied, but encouraging. A licence was finally granted for the application of formic acid by means of the 'Illertissen Mite Plate' (or IMP in Germany) above the brood nest. It is a sheet of soft card board (beer mat type) soaked with 14.2gms of pure formic acid (about 20 ml of 85% formic acid). Placed above the brood nest immediately after the removal of the honey crop, the acid vapour kills many mites on bees.

The results had often been disappointing, especially so when native-type bees maintained deep honey rims above the brood. This honey crown has been shown to be an aerodynamic barrier against losses of warmth and humidity from the brood nest (Mobus 1977). The narrow beeway between the sealed honey crown above the brood prevented the fumes of reaching the young bees in the brood nest, and concentrations remained too low to be effective. New experiments have shown that mats with formic acid placed below the brood nest are far more efficient in reaching killing concentrations in that very area where most of the infected bees congregate: the brood nest.

The delayed, continuing 'raining down'

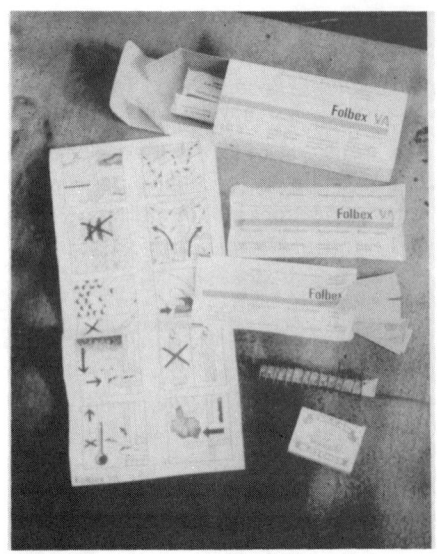

FIG 29 *A pack of Folbex VA with instructions. An applicator for the fumigation strip made from wire is shown at bottom right.*

FIG 30 *Fumigation with Folbex VA. After flying has stopped, the strip is lit and pushed through the entrance with an applicator. The entrance is closed for an hour.*

of varroa mites after an application of formic acid raised hopes that the material can penetrate through the cappings of infested brood cells. Experiments (Adelt & Kimmich 1987) have confirmed the deep-reaching action and have shown that brood frames (without bees) can be treated effectively in boxes constructed of insulating material. In a chamber for 10 frames (about Langstroth dimension), 30 ml of 60% formic acid on soft card board (40 cm x 11 cm) above or below the frames, and an exposure time of 90 minutes, have proved sufficient to kill 80-95% of the parasites in the cells. Open brood is hardly damaged, and only bees emerging during fumigation may be killed. Drone comb needs slightly longer time and greater concentrations. The amount of brood and weight of open honey stores (absorbing much of the acid) can influence the amount of acid required. Cardboard mats should be soaked the day before application and kept in airtight plastic bags for even distribution of the acid. Temperatures in the fumigation box should not drop below 25°C. It has been found that as long as external temperatures lie above 20°C (70°F), no heating element is required. The box should have enough space (50 mm; 2") above and below the frames for free diffusion of the vapour.

The use of 'natural' fumes from essential oils also raised many hopes. Some workers hoped that the smell might confuse varroa mites in their choice of cells, and the idea of the treatment of colonies against the tracheal mite with fumes of methyl salicilate had given them encouragement. Others believed in the curative aspects of the chemicals used in this version of 'aromatherapy' for bees. Thymol, menthol, oil of cloves or marjoram and many other 'smelly' materials were tried with variable rates of success. Nearly all vapours affected the flavour of honey and perforce, are illegal administrations.

Anesthetising vapours such as ethyl alcohol, chloroform and ether are unsuitable, the last two knocking out all bees and making them vomit the contents of honey sacs. This can result in the bees drowning in nectar, honey or sugar syrup. The search continues for some gaseous agent which penetrates the minute body of the mite before it can affect or kill the very much larger bee.

Chemical Aerosols and Fumes

It is in this area that most of the earlier successes were achieved. Chemicals were often sprinkled on glowing embers in the smoker, and the resulting fumes were forced into the hive. Phenothiazine had always been used in this haphazard way; now there are easier ways of administering precise amounts of an effective chemical. Nowadays there may be fumigating strips or pills which consist of a combustible material and an oxidising agent - as well as the varroacidal chemical. The carrier is often just a strip of soft blotting paper soaked with a small amount of saltpetre solution (sodium nitrate) to sustain slow burning. Folbex (old formulation, used against acarine disease) was one example, Folbex VA is the new substitute. The active ingredient of Folbex VA is bromopropylate. This chemical is very fat soluble and wax absorbs the drug. The constant applications - 4 fumigations are recommended at intervals of 4 days (annually!). will, over the years, cause a build up of the chemical residue in wax. Therefore even the finest particles of cappings wax must be removed from honeys, and the production of section honey or comb honey must be restricted

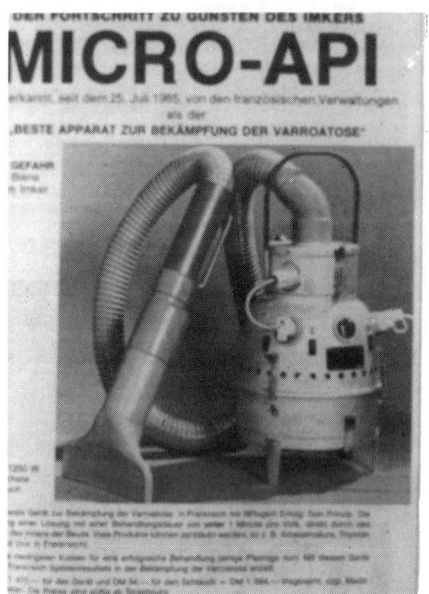

FIG 31 *Advert in the popular bee press showing fume or aerosol generator for use with Amitraz. (Taken from Adiz)*

to non-users of Folbex VA - and other chemicals. In many European countries it is officially approved for treatment (against varroa and acarine disease) when no brood is present in autumn and before winter temperatures enforce clustering. At the recommended dosage bees tolerate Folbex very well, and even four times the normal dose caused no losses. Broodless colonies, natural and shaken swarms are treated efficiently and successfully, but any mites in brood cells are unaffected.

Amitraz, originally a horticultural acaricide, is used in many countries. Poland has mixed it with combustible materials and an oxidising agent and formed it into pills for use in the smoker, France injected it, atomised with steam (but temperatures lowered to 40°C; 100°F) as an aerosol into the hives, and special steam generators quickly appeared on offer in the bee press. Yugoslavia suggests the home-manufacture of fumigating strips (see above) treated with 0.5ml of 25% Tactic (the trade name for the active chemical in a commercially available form). Rumania has abandoned its formerly highly favoured powdering of bees with Sineacar in favour of Amitraz on a fumigating strip.

At this moment - but things are changing rapidly - Amitraz appears in more combinations and concoctions than any other drug. This fact may mislead us into feeling safe when using some form of Amitraz, but its safety record is under examination in California, and Germany has also not given it approval for use in colonies of honeybees. Old reports of its carcinogenic character are hard to live down, and new, intensified research may yet bring bad, or good news. In the form of Tactic it has been in garden shops for some time. Special formulations for use in apiculture are usually very expensive.

Fumigation with Amitraz lends itself to mechanisation or, at least, to simplified mass treatment. The French steam generator for use with Amitraz, complete with built-in fan, should make treatment of many hives an easy job. In France many local beekeeping associations have invested in a generator and then lend it out to members. The steam generator can be run on bottled gas, the fan needs car batteries for use in out-apiaries. Treatment is best applied in the evenings when most flying bees are at home. Its use should also be restricted to the periods when no brood is present in the hive and when no honey flow is expected. Often the beekeepers themselves breathe in a greater amount of the chemical than the colonies to be treated. Masks should be worn by all who use atomised chemicals or aerosols. This is a danger not sufficiently appreciated by beekeepers, but warnings should appear repeatedly in the bee press.

Micro-fumigants should also come into the picture in this chapter. They are highly active chemicals with such a low vapour pressure, that the amount of the chemical in air is possibly at the Nano level. (Vapona strips belong into this class, but let no one play around with that organophosphorous compound!)

Some of our safest and best insecticidal chemicals belong into the family of the natural pyrethroids. A new, synthesised member of the family is Fluvalinate and shows great promise as a handy varroacide. Apistan is a strip of plastic which is saturated with the chemical. It must be mentioned in this group, even though most workers declare it to be a systemic poison, after its absorption by bees by contact with the strips. Prof. Roger Morse, for example, writes the following in the December 1987 issue of "Gleanings in Bee Culture": The preferred formulation is that Fluvalinate is impregnated into plastic strips that are placed in the broodnest. Bees walk over the strips and pick up the chemical. The mites continue to be killed as they emerge from their cells".

This implies that bees absorb the chemical through their sticky pads (pulvilli) on their feet, and that varroa mites taking a blood meal are then poisoned. Yet its fast and thorough action on all mites as they emerge from the cells raises many doubts. Because not all bees make direct contact with the strips, it must be passed from bee to bee by social feeding. This is unlikely, because the honey stomach from which social food exchanges take place, is lined with chitin and resists the rapid exchange of chemicals. Even water from the nectar is not absorbed by the surrounding tissues. Yet the chemical is supposed to pass in a counterflow from the blood into the contents of the honey

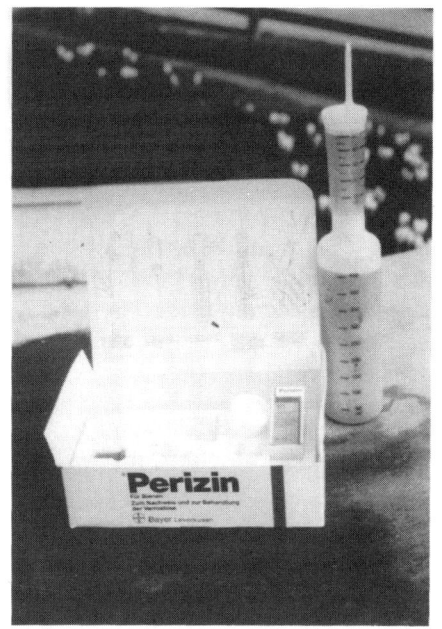

FIG 32 *Perizin outfit. The clever applicator permits the rapid dosing of up to 20 colonies of standard strength with 50 ml medicated Perizin solution (water or syrup). Weaker colonies are given half the quantity.*

stomach! So Apistan must belong to the category of micro-fumigants and in this group.

After the chemical's initial successes in various countries, one firm, Zoecon, a subsidiary of Sandoz, Texas, succeeded in obtaining a provisional licence for the production and sale of plastic strips which had been impregnated with a measured amount (0.80 g) of the chemical. This became available under the trade name Apistan and has shown great promise. Strong colonies needed two of these strips which were left in the hive for 6 weeks. After that period the strength of the poison is so low that mites could develop resistance to the chemical. This must be avoided at all cost, and all strips should be removed from the hive after 6 weeks.

French beekeepers certainly prefer this treatment to getting their lungs filled with dusts and poisons. The price of Apistan being rather high, some of them manufacture their own strips by soaking strips of cloth or soft wood in a 2.5% solution of Klartan, or spray a watery solution over bees and combs (see **"Fighting Varroa in Southern France"**).

However, we should also take heed of the warning given to German beekeepers "who want to play researchers", that fluvalinate could have a neuro-toxic effect in humans. So it is best to isue a warning here: **Handle all chemicals, including Apistan strips with great care - and gloves!** No contamination of honeys with fluvalinate has been reported.

Sprays

Water soluble materials can easily be applied to bees in the form of sprays. Non-soluble drugs can be emulsified if a suitable, harmless emulsifying agent is incorporated. A fine spray can then be applied to every bee and comb in the colony. The greatest danger to bees during cool weather is the thermal shock and the loss of heat from the evaporation of the water. Applications should therefore only be made when temperatures are warm and the air is drying.

New chemicals turn up in this category, some of them being very simple, indeed, natural substances. Acetic acid was probably one of the first to be tested, but this soon gave way to lactic acid. Commercially available lactic acid (used in the food industry) can be diluted to a 13 - 15% solution, and this is sprayed over all bees on the combs, wetting them all over. To bees and humans it is fairly harmless and is found naturally in all meat, cheese and yoghurt. Reports speak of 68 - 97% success rates on adult bees. Varroa in cells are not affected.

Oxalic acid is preferred in by some workers in Russia, and Japan has also tested it successfully. In this case a 2.7% solution of oxalic acid has been applied to all bees on comb, and this reduced the varroa population considerably. Klartan (fluvalinate) is also used as a watery spray by some beekeepers in France.

Chemical Dusts and Powders

They are usually made up by mixing a larger quantity of a neutral carrier base and small dose of a powerful chemical. Many materials have been used as the carrier base, and powdered gypsum, finely ground marble, baking flour or dextrose powder rank among the 'inert ingredients'. The chemical used as the varroacidal drug varied from country to country. Greece preferred malathion at 0.05 g/kg of base. Rumania formerly used a mixture of chloropropylate and bromopropylate under the name of Sineacar (forte), and selling it by the tonne for mixing with flour or similar dusts. As in all other cases, mites in the brood cells are not affected, and the broodless period, after all chances of a honey crop are over, is recommended as the best time for applications.

Newest reports point to the fact that the efficiency of all chemical dusts may partially rely on the mechanical action of dust contamination on the mite's sticky pads. This aspect has been mentioned above (see Physical and Mechanical Methods). Maybe the Greeks and Rumanians could have left the chemicals out of their 'successful' formulations and could have produced safer, healthier honey?

Control with Systemic Acaricides

This would be the easiest way of killing mites. Bees are fed a substance which is harmless to them but which, after entering the haemolymph, kills all mites when they are taking a feed of blood. Sugar syrup or water suggests itself as the easiest diluting carrier for medication, and social feeding, well researched and found to be highly efficient among social bees, distributes the drug to all bees in the hive. The problem is finding the right drug and avoiding all contamination with honey, however 'harmless' the poison itself may be.

A chemical, known at first as K-79, raised great hopes in Germany in the early 80s. A minute quantity was dissolved in water and 50 ml of the solution were applied twice at four day interval, when the colony was without brood. Success rate was better than 95%, often as high as 98% or a complete 'cure'. Strenuous efforts to obtain approval by the authorities failed; the substance was known to be carcinogenic, and the chemical had already been banned as an acaricide for horticultural use. Even though the small quantity of 15 microgramme in 50 ml water would never have reached next summer's honey crop, no approval for its use in honeybee colonies was given.

The chemical industry has investigated other compounds since then. Some are known to have acaricidal properties and are of relatively low toxicity to bees and humans. Two of these are at present showing promise; one, Perizin (Bayer) has gained approval in Germany in spite of that country's stringent food laws. The other one, Apitol (Ciba-Geigy), is awaiting official approval.

Perizin contains an organophosphorous compound, and is therefore related to the wicked nerve gases. That alone sends shivers down our spine. On the other hand, the same formulation, under a different name, is used on American farms against ticks and mites on cattle. The Food and Drug laws of America have set an upper limit as high as 0.1 ppm in meat for human consumption. For use in bee colonies, a small quantity of 32 mg of Perizin is emulsified with water or sugar solution and is dribbled over the bees of a broodless colony. Temperatures should not be below 10°C (50°F), and bees tolerate it better above 15°C (60°F). Low relative humidities permit wet bees to dry rapidly and suffer less shock through wetting.

The firm of Bayer has devised a clever device of metering quantities of the emulsion, so that small colonies and nuclei receive half the amount or less, and full strength stocks can be given the full 50 ml. Twenty stocks can be medicated with one full applicator. The work is simple and quick, and many colonies can be treated in a short time. By using larger reservoirs or applicators the process can be speeded up in order to cope effortlessly with a larger number of colonies in one apiary.

Perizin is not completely harmless to bees! Many bees exposed to the full dose may die before they pass some of it on to other bees in diluted form. Provided the dosage had been correctly measured and applied during mild weather, the loss of a few, occasionally up to 200 bees, must be disregarded. It is better to lose some bees than a whole colony. Where beekeepers took fright and reduced the dosage after seeing bees dying, the effectiveness of Perizin was greatly reduced, thus endangering the stock as well as running the risk of producing a resistant strain of varroa. On the other hand, in one bee-compatibility trial with the drug in Scotland, Perizin was administered on a cool day (7°C; 47°F) to a colony of moderate strength (no varroa). It received a double dose experimentally and lost in excess of 1000 bees - yet it pulled through the winter in great health. Perizin is said to have a knock-down effect as well as a delayed action, as the dropping down of varroa mites continues over several days. Mites emerging from cells in the following 3 - 5 days may still receive a killing dose when imbibing the blood of adult bees. (See also **"Fighting Varroa in Southern France"**)

The price of Perizin is high, but beekeepers have already discovered that the same drug is available at a fraction of the cost for veterinary purposes in apparently the same formulation. Of course, the mixing of a home-made solution with insufficient precision could lead to miscalculations and heavy losses. Perizin has a strong smell, and every effort must be made to ensure that the treatment is administered only in autumn when the colony is without brood and there is not the slightest chance of contamination of next year's honey crop. The active compound in Perizin is fat soluble and will therefore get into the wax. All wax, even the finest particles of cappings, must be removed from honey by filtration before being offered for human consumption. Colonies which are worked for sections should be given other treatment.

Ciba-Geigy of Folbex VA fame hopes to have Apitol, a new varroacide, available shortly. This too has systemic action, and is fed to colonies in autumn in a quantity of sugar syrup, and it shows promise. Information about dosage, frequency of treatments needed, and its effectiveness is still lacking at this moment. Official approval for its use with bees is expected shortly in Germany. The formula suggests that it is water soluble, and that wax will not be contaminated.

A full list of chemicals, their common names and trade names is included in the appendix.

Residual Contamination of Honey Crops

To our customers of honey, the word 'Chemicals' spells contamination, residue, pollution and all things nasty. On the other hand, we do claim that honey is some of the purest, healthiest food. The use of chemicals, even in minute quantities, spells out a problem, because modern technology has perfected the methods of detecting contaminants to such a fine degree, that single parts per million or billion (ppm, ppb) can be traced and identified, however harmless they may be at such low levels. Our hive products, and they do not only comprise honey, but also royal jelly and propolis, must remain beyond suspicion. When the press hounds find out and spread the sensational news that the dangerous, destructive mite is being fought by means of 'chemicals' on all fronts, and that local honey therefore could be suspect, the local Health food shops will refuse to buy 'British Honey' and advertise their foreign honeys as clean, healthy, organic and uncontaminated - even though the country of origin may have less stringent regulations - and beekeepers there get away with murder. Our products must remain above suspicion, and we must limit all applications of chemotherapy against varroa to a time when no drug can find its way into honey, and to a level which ensures a kill, but not a contamination.

Resistance to Chemicals

Bugs and wee beasties have the tendency to become resistant to a chemical used frequently to suppress them. Strains of bacteria have become resistant to antibiotics, some mosquitoes shrug their shoulders when encountering DDT, and even rats and mice can tolerate Warfarin now. Varroa can also become resistant to chemicals, and some reports, unconfirmed, state that heavy losses of colonies have occurred in France after continued use of Amitraz. We must therefore have a battery of different chemical weapons at the ready at all times. It may even be best if we ring the changes at times.

In the case of the simplest, 'natural' chemicals, such as formic acid, oxalic acid or lactic acid, the chances of resistance are least likely to come about. In one respect we can be certain: varroa is least likely to lose its sticky pads and

The New Varroa Handbook

grow claws with which to grip to the bee's hairs. When all else fails, we will always have dusts and powders to reach for. It is strange, but Dr. J. Rennie recommended the flouring of bees in 1922 in order to prevent the tracheal mite from being able to migrate to new spiracles! Nothing is new under the sun, but it took time before we became aware of this simple means of controlling varroa. The application of fine dusts can probably still help when varroa has become resistant to all other chemicals in use today.

Last Words

There is another point which we must never forget. The happy-go-lucky amateur beekeeper will have to mend his or her ways. He must be aware of the presence of varroa and of the danger which the mite presents to the survival of his stocks. Lackadaisical application of chemicals, either in terms of choice or strength, of time or ambient temperatures, will lead to the collapse of colonies after a few years. Greater care and attention to detail will be required from now on. More lectures, teaching and sheaves of advisory leaflets will not change matters; greater learning is needed so that it can lead to better understanding. Maybe this little booklet will make a small contribution towards preparing the beekeeping industry for the coming of *Varroa jacobsoni* in the your hives.

THERAPEUTICAL CHEMICALS AND TRADE NAMES   1.9

Many chemicals have been tried and tested in the figtht against Varroa in the course of the past 15 years. Some of the drugs were used on their own, others were applied as mixed formulations with other active substances in the hope of making certain of increasing the effect of the other.

The following list contains many, but not all of the drugs and chemicals, their chemical names, their registered trade names and, where applicable, the trade names of patent medicines (veterinary) which contain the active ingredient in the same or similar formulation. The sign =~ does not intend to convey that the concentrations of active ingredients are equivalent.

The list makes no claim to be comprehensive, indeed, it makes no effort to list home-made concoctions used in desparation by beekeepers, e.g. spraying with copper sulphate (in the wine-growing districts of France) or the 'dynamic approach' (a homeopathic concoction of charred varroa remains which is sprayed over the bees and brood. That was guaranteed to work by my informant).

ACAREX =~ Dinobuton=isopropyl 2,4-dinitro-6-sec-.butyl-phenyl carbonate

ACAROL =~ Bromo-propylate =~ Di-bromobenzilate =~ FOLBEX VA =~Neron =~Isopropyl-4,4'dibromobenzilate esther

ACRICID =~ Binapacryl =~ 2-sec-butyl-2,4-dinitro-6-butyl-phenyl-3,3-dimethylacrylate

AKAR =~ Chlorobenzilate =~ Chloropropylate =~ Ethylene-dichloro-benzilate =~ FOLBEX (old) =~ ROSPIN =~ 4,4'dichlorobenzylicacid ethyl ester

AMITRAZ =~ FUMILAT =~ HEMOVAR =~ MITABAN =~ TACTIC =~ TCL =~ TETRAMIX =~ TRIAZID =~ VARAMIT =~ VAROCID =~ VARRESCENCE =~ ANTI-VARROA SCHERING =~ N-methyl-bis-(2,4-xylyliminomethyl)-amine =~ 1,5-di-(2-4-dimethylphenyl)-3-methyl-1,3,5-triazapenta-1,4 diene

ANTIVARROA SCHERING see AMITRAZ

APIACARIDIUM =~ Malathion + TEDION

APISTAN =~ Fluvalinate, SPUR, KLARTAN, a synth.pyrethroid

The New Varroa Handbook

APITOL =~ 2(2,4-Dimethylphenyl-amino-3-methyl)4-thazoline-hydrochloride

ARAMITE =~ Sulphurous acid 2-Chloroethyl-2-[4-(1,1-dimethylethyl)phenoxy]-1-methylethyl ester

ASUNTOL =~ CO-RAL =~ PERIZIN =~ RESITOX =~ Coumaphos =~ 3-Cloro-4-methylumbelliferon-0,0'-diethylthiophosphate =~ 0,0'-Diethyl-0"-(chloro-4-methyl-7-coumarinyl)-thiophospate

Binapacryl - see ACRICID

Bromopropylate - see ACAROL, (Folbex VA)

Chinomethionate =~ MORESTAN =~ OXYTHIOQUINOX =~ 6-Methyl-chinoxylin-2,3-dithiol-cyclocarbonate

Chlorobenzilate - see AKAR, FOLBEX (old), ROSPIN

Chlorodifon - see DANICAROBA

Chlorodimeform - see K79

Chlorophenamidine - see K79

Chloropropylate - see AKAR

CO-RAL - see ASUNTOL, PERIZIN, RESISTOX

Coumaphos (Coumafos) - see ASUNTOL, PERIZIN

Cyhexatin - see PLICTRAN

DANICAROBA =~ POLYACARITOX =~ TEDION =~ Tetradifon =~ 2,4,5,4'-Tetrachloro diphenyl sulfone

Dekachlor (Decachlor) =~ HRS =~ PENTAK =~ Dienochlor =~ Bis-(pentachloro-cyclopentadien-2,4)-yl)

DIAGVAR =~ SINEACAR =~ Chloropropylate + Bromopropylate

Dibromobenzilate - see ACAROL; FOLBEX VA, NEORON

Dichlorobenzene - no information

DICOFOL =~ KELTHANE =~ 1,1,1-Trichloro-2,2-bis(-4-chlorophenyl)-ethanol
Dienchlor - see Decachlor

Ethylenedichlorobenzilate - see ACAR, FOLBEX (old), ROSPIN

FLUVALINATE =~ SPUR, KLARTAN, APISTAN (synth. Pyrethroid)

FOLBEX - see AKAR, ROSPIN, Dichlorobenzilate

FOLBEX VA (also FOLBEX VA NEU) see ACAROL; NEORON

Formic acid - Natural material, used as IMP (Illertisser Milben Platte)

FUMILAT AMITRAZ; HEMOVAR, MITABAN, TACTIC, TCL, TETRAMIX, TRIAZID, VARAMIT, VAROCID, VARRESCENCE, ANTIVARROA SCHERING

GALECRON - see Chlorodimeform; K79

HEMOVAR - see AMITRAZ

HRS - see Dekachlor; PENTAC

IMP =~ Illertisser Milben Platte; approved method of application of formic acid

K-79 - see Chlorodimeform; GALECRON

KELTHANE =~ DICOFOL

KLARTAN =~ FLUVALINATE, SPUR, APISTAN, synth. pyrethroid

Lactic acid - natural material

MALATHION =~ 0,0-Dimethyl-S-1,2-bis(4-chlorophenyl)-ethanol

Marjoram - natural material

Menthol - natural material

MICAZIN - no information

MILBEX - no information

MITABAN - see AMITRAZ

The New Varroa Handbook

MORESTAN - see Chinomethionate

NEORON - see ACAROL

OMITE (OMIT) =~ Propargit =~ 2-(p-tert.-butylphenoxy)-Cyclohexyl-2-propynylsulfit

Oxalic acid - natural material

OXYTHIOQUINOX - see Chinomethionate

PENTAC - see Decachlor

PERIZIN - see ASUNTOL; CO-RAL, RESISTOX, Coumaphos

Phenothiazine =~ VARITAN =~ VARROATIN =~ Thiodiphenylamine =~ Dibenzo-1,4-thiazine =~ Dicresylether-methyl-carbanimic acid

PLICTRAN - see Cyhexatin

POLYACARITOX - see DANICAROBA; TEDION

PROPARGIT - see OMITE (OMIT)

RESITOX - see ASUNTOL; PERIZIN

ROSPIN - see AKAR; FOLBEX (old)

SINEACAR - see DIAGVAR

SPRUZIT (Pyrethroid)

SPUR =~ APISTAN, KLARTAN, Fluvalinate (synth. Pyrethroid)

TACTIC - see AMITRAZ

TCL - see AMITRAZ

TEDION see DANICAROBA

Thymol - natural material

Tetradifon- see DANICAROBA

TETRAMIX - see AMITRAZ

Thiodiphenylamine - see Phenothiazine

TRIAZID - see AMITRAZ

VARAMIT - see AMITRAZ

VARITAN - see Phenothiazine

VARIZID - see AMITRAZ

VARRESCENCE - see AMITRAZ

VARROASIN =~ Phenothiazine + 2,4,6-Trimethyl-1,3,5-trioxan

VARROATIN - see Phenothiazine

VARROSTAN - no information

VAVAPIN =~ Phenothiazine + Malathion

1.10 BIBLIOGRAPHY

ADAS Leaflet 936, various reprints; Use of Tobacco Smoke

Adelt, B and Kimmich, K (1986): Die Wirkung der Ameisensaure in die verdeckelte Brut. (Action of formic acid in capped brood cells) ADIZ 20/12; pp 382 - 385

Ball, B. (1983) Der Zusammenhang zwischen *Varroa jacobsoni* und Viruskrankheiten der Honigbiene (Relationship between *V. jacobsoni* and viral disease in honeybees) (Transl. Koeniger)
ADIZ 17 (6); 177 - 179

Holler, G (1987) Praxisversuche mit PERIZIN zur Bekampfung von Varroatose; (Field Trials with PERIZIN in the Control of Varroasis) ADIZ 1987 21/1 p23-24

Holzer, J. (1987) Zelt und Freilandversuche mit PERIZIN an Bienenvolkern mit Brut. (Tent and Field Trials, Using PERIZIN on Colonies with Brood) ADIZ 1987, 21/4; pp 121 - 122

Huttinger et al (1981), Report, Apimondia Symposium on Varroa

Koeniger et al, Interim report on the use of lactic acid in the treatment of Varroasis. ADIZ 1983/7

Liebig, G (1986): Ameisensaure, FOLBEX VA, PERIZIN; - ein Vergleich; (Formic acid, FOLBEX VA, PERIZIN; Comparative trials)
ADIZ 20/8 pp 254 -256

Maul, V. 1988, Report: Ist die Nachbewertung der Herbstbehandlung zur Beurteilung von Zuchtbestanden hinsichtlich unterschiedlicher Resistence geeignet? (Is is possible to assess differences of varroa-resistance by evaluating the effect of autumn treatment of breeder colonies?) Freiburg, November 1988

Mobus, B (1975) The Ideal Brood Frame, Gleanings, Vol.103; p 355

Mobus, B. (1979) Brood Initiation in the Winter Cluster, Report, APIMONDIA Congress, Athens 1979, pp 244-248

Muller, M (1987) Befallsentwicklung der Milbe *Varroa jacobsoni* in den Wintermonaten, (Development of Varroa Populations during Winter Months, ADIZ 1987, 21/1; pp 15-22

Muller, M (1987) Varroa Bekampfung durch Einsatz der Ameisensaure von unten, (Controlling Varroa by the use of formic acid below the brood nest), ADIZ 21/7/1987 pp 216 - 218

Ramirez, W.B., Escuela de Fitotecnia (1987), (Translation by Schmadalla, ADIZ 1988/2 pp 62 - 63), Brazilian

Congress 1987: Newsletter for beekeepers in tropical and subtropical countries; pp 10-11

Rehm, S (1988), Report: Reihenfolge und Entwicklungsdauer der mannlichen und weiblichen Nachkommen von *Varroa jacobsoni* in der Arbeiterinnenbrut von *Apis mellifera carnica*; (Sequence and times of development of *Varroa jacobsoni* in worker cells of *Apis mellifera carnica*) Freiburg, November 1988

Ritter, W. et al (1987) Schwarmverhinderung mit Hilfe des Bannwabenverfahrens, (Varroa control, combined with swarm control by means of 'Varroa bait comb' and the 'Artificial Swarm')
ADIZ 87/6 p 184-185

Ritter, W. et al (1987) Die Bildung von Ablegern aus behandelten Kunstschwarmen von FOLBEX VA, PERIZIN and Bannwaben, (Formation of nuclei from shook swarms treated with FOLBEX VA, PERIZIN and Varroa bait combs) ADIZ 1987 21/7 pp 219 - 220

Ritter, W., (1986); Lecture, "Varroa, The Mite's Biology"; Craibstone, Aberdeen, Scotland, September 1988

Ritter, W. (1989), Lecture; "Experiments during the development and trial of PERIZIN, a systemically acting medication for fighting Varroasis in honeybees", Aberdeen, Scotland. September 1986

Ritter, W (1986): Varroatosebekampfung: Praktische Hinweise aus dem Tierhygienischen Institut in Freiburg, (Controlling Varroasis: Practical hints) ADIZ 20/10 pp 334 - 335

Sabolic, M. and Unglaub, W. Die Belastung von Honig mit PERIZIN (Contamination of Honey with PERIZIN), ADIZ 1987, 21/2 p 45

Schneider, P., et al (1988); Die Folgen eines unterschiedlich hohen Varroa-Befalls wahrend der Puppenentwicklung auf die erwachsene Bienen, (The consequences of variable degrees of parasitation by Varroa on pupal development and the adult bee) Adiz 22 1 / 2; 16-18 ; 54-56

Schulz, A.E. (1984) Reproduction and population dynamics of the parasitic mite *V. jacobsoni* and its dependence on the brooding cycle of its host *A. mellifera*; Apidologie 15 (4), pp 401-420

Strick, H. and Madel, G. Varroatose und bakterielle Sekundarseuchen. (Varroasis and secondary bacterial diseases) ADIZ 1986, 20/10 pp 321 - 324

Weiss, J. (1987) Mit Milchsaure gegen die Varroa-Milbe Control of TactiVarroa with Lactic Acid) ADIZ 1987/8 August, p 258 - 262)

Wienands, A., Madel, G (1987) Bienen, Blut und Parasiten, (Bees, Blood and Parasites), Adiz 1987 21/1 pp 8-10)

2 TREATING VARROA IN THE SOUTH OF FRANCE

I am now living here in the South of France, where the wine is good and cheap, where this year (1991) the drought lasted only 20 weeks, and where often the day-time temperatures of summer occasionally soared above those of the brood nest. I am retired now and had intended to give up writing to the bee press, but recent experiences, as well as some arm-twisting, made me change my mind.

Yes, I am still a beekeeper - of sorts. I own one hive at the moment. After some hesitation, I bought it for £50 around mid-October last year (1990). The previous owner had kept them in the hills at an elevation of 3000ft for the past 10 years, and told me that the bees were of the 'little black mountain bee' strain. Native, or near-native bees; just what I had been looking for. The hive was a Dadant deep, easily 100 lb in weight, and a shallow super was thrown into the bargain.

Once the bees had settled, I gave them a cursory (they were not very nice when disturbed) inspection and found two strips of cloth, 1" x 6" long, under the crown board. They were solidly coated with propolis. The strips had been a home-made method of varroa control applied over the tops of the frames. But when? One, two, or three years ago? Still, I hoped that the varroa population would be small, especially as I found only 20 parasites on the floor board. This had never been removed for many years, nor had the frames. The low number of parasites on the floor made me relax my vigilance.

Spring brought some problems, but they had nothing to do with varroa - nor with the AFB I discovered. On many days I seemed to lose a lot of bees of nursing age. Well, the bees had been collecting pollen nearly every day during winter. Brood rearing seemed continuous. Flight only stopped when the weather turned too windy or to cool. Each first flying day after a period of confinement, many, sometimes hundreds of young bees walked away from the hive with swollen abdomens. The explanation is simple. Here in France it would be called 'Mal de Mai', but its cause would be a different story from mine.

In spite of water being available nearby, workers had been unable to collect any of it on windy days when the 'Tramontane', the local 'Mistral', blows for days. The young nurse bees, who eat lots of pollen in order to produce brood food, had been unable to replenish the loss of water needed for its elaboration. In order to maintain blood concentrations at an equilibrium, moisture had then been resorbed from the rectum, and the contents of that part of the bowels dried up.

Bernhard Mobus

Constipated, with a solid plug of pollen debris blocking the exit, the bees suffered badly when fresh water and nectar became available again. In spite of being bloated now with a watery burden, bees could not evacuate any longer.

My bees seemed an idle lot compared with my Scottish strain, and no honey was produced during spring and the height of summer - in spite of living in a bee's paradise. Mimosas were in full bloom by February, and perfumed garlands of wisteria trailed over the pergolas in March. The hill-sides were white with mediterranean heather at the same time, and the first poppies flowered in April by the wayside. By mid-May some fields were white with clover, but the conditions had possibly become too dry for nectar secretion. Just the same, honey began to flow during July and early August, when bees worked steadily in spite of the prolonged drought. By mid-August I harvested 15 kilos (33lb) of lovely honeydew honey called here, politely, 'miel de montagne 'or' chene vert' (evergreen oak). Pale brown and strongly Virol-flavoured, it tasted beautiful, although in Britain it would hardly gain prizes at a honey show. But dark honeys carry a premium here in France. Well, the sun shone and there was honey for breakfast; and varroa seemed the least and the last of my problems.

After believing that I had got away with the problem of varroa, and hoping that somehow the high temperatures had inhibited the brooding of bees or had killed many mites, I sat back, twiddled my thumbs and enjoyed the sun, the mountains and the sea. By now it was just too hot to work on the house or in the garden. Just the same, one day in late August, when I was watching my bees and talking to them (in English, my French is poor), I saw two of them perform a violent grooming dance on the flight board. I became suspicious. Picking them up and examining them carefully, I found one varroa mite clinging to each bee between the thorax and the abdomen. Although the grooming dance had elicited assistance from other bees, the help given by the friendly sisters had not been very effective. The mites had remained on their victims.

This discovery made me decide on a diagnostic varroa control treatment in order to find out the full extent of the problem I had hoped I never had - yet. At first I hesitated a little over the decision to treat the colony with a varroacidal poison. After all, the instructions on the package of the drug Perizin (which I still had in stock after trying it out in Scotland) warned you to administer the drug ONLY during a broodless period. But the way things are here, bees were collecting pollen by the bucketful nearly every month of the year and brood rearing seemed never out of season. Well, I could not wait and let the bees suffer for ever, and so, for a start, a new floor was placed underneath the hive. Its screened plywood floor, below a slatted 'Killion' floor, could be withdrawn from the back with little disturbance to the colony.

Finally, taking the bit into my mouth, I applied a full dose of Perizin (Bayer) on September 10th in mid-afternoon. By the evening of that 'Day Zero' I took the first count and found 69 varroa mites on the floor. Some seemed very pale, like straw, others were darker, but not of the deep brown colour of the mature mites. The rest, all 58 of them, I simply called 'mature'. Later I was to discover that even some of these dark brown mites of the first kill were probably still youngsters in an advanced stage of maturation. Allthough fully coloured, they were young mites just itching for reproduction (all female varroa mites are fertilised before leaving the cell). All shades of pale straw, light and medium brown were lumped together after that, and their presence in the sample (**Table 2**) simply reflects that the colony is still rearing brood.

The first count on Day 1 stunned me; I had not realised that things were that bad (but worse was to come). There were 230 mites on the floor. I made two further collections that day, and 26 and 19 more mites were counted respectively. The total of Day 0 and Day 1, after one application of the drug, amounted to 344 varroa mites. Foolishly I trusted that the drug had knocked down every mite which had not been inside the cells. Because the advertising claims that the killing effect of the systemically acting medicine persists for several days, I also hoped that the count of the following days would represent the full daily kill of all mother mites and their brood, as they emerged, together with their victims,

from the cells. The size of the daily kill left me confident that this was so. Yet when a new dose of Perizin was given on Day 6 in order to top up the concentration of the poison in circulation, the new count was far too high to have been just one day's emergence of mites. Just the same, the steady kill continued after that and made me happy, little realising that the deaths were daily losses from 'natural causes'.

Because bees require 12 - 13 days development under the cappings, Day 14 was expected to show a decline in the numbers of mites found on the floor each morning. Yet the expected drop did not materialise. Instead, on Day 18 the daily roll call increased and was was higher than ever before. So, after a morning count of 114 mites, I made an efficiency check on Perizin by fumigating the colony with two strips of Folbex VA (Ciba-Geigy). A frighteningly large number of mites were knocked out: 615 mites, and that on a day when 114 varroa had been counted already. This new kill consisted mainly of old, dark mites with few pale and medium brown mites. They must have been older mites which had escaped being poisoned by Perizin and had stayed alive. Folbex VA caught them before they could enter cells for the first time.

A closer examination with a stereo-microscope (25X magnification) had shown that the 'pale straws' were often only shells without legs or bodies. It appears that when the mother mite is carrying on laying eggs, the varroa mites hatching from the last eggs cannot develop fully. Some immatures have reached the soft and vulnerable deuteronymphal stage. It appears that about 20% of the varroa mites emerging are too young and will probably never make the grade. Several badly mauled left-overs of protonymphs were also discovered among the debris. Although, theoretically at least, bees could have damaged the mites before they fell through the screen, the number of badly mutilated varroas permits the interpretation that the first-born varroa mites probably turn on their hapless sisters when the chitin of the maturing baby bee is getting too hard and cannibalise them as an easy prey. Just the same, it is immaterial whether the varroa grows up or dies; we must never forget that the palest mites and even the protonymphs have been draining the life-blood of the developing baby bee and had damaged it for life.

Although the daily 'kill' of varroa mites continued for many days, I should have realised by now that that number could represent the daily death rate for natural reasons. However good the chemical may be during late autumn or early winter, **Table 1** shows clearly that Perizin is not an efficient drug when the colony is of summer strength and is still brooding and working. Not all worker bees or nurse bees receive any - or consume enough of the drug - to reach or maintain a killing concentration of Perizin in the haemolymph. The division of labour into groups of foragers rarely making contact with nurse bees and baby bees prevented the even distribution of the drug. Here in France, many varroa mites survive to infect more cells. When a swarm is treated, or a winter cluster without brood, it may well be a different story.

The ever-increasing daily count after Day 15 proved to me that the kill achieved with Perizin had no effect on the population of the parasite under my conditions. Folbex VA, on the other hand, had the better knock-down effect. Yet, being a smoke, it lacks staying power and cannot penetrate through the cappings. Therefore, this form of control by fumigation should also be applied ONLY during a broodless period when all mites are outwith the cells.

Perizin is not without harm to the bees. About 100 bees seemed to die after each application of the correct dose. Under the conditions prevailing here in the South of France even Folbex VA caused some loss of life. Maybe the volume of the brood box of Dadant dimensions, plus a shallow, did not warrant two strips of the fumigant during the warm weather (20°C; 68°F).

Of course, any minor loss of bee life occasioned by an efficient drug must be borne with fortitude; it is far better than seeing hundreds of crippled bees thrown out of a hive which is in a state of total collapse. In stark contrast with other bee diseases, the collapse due to varroa usually happens around July, August or September. At that time the large population of mature mites dives into the worker cells of a shrinking brood nest, two, three at a time, for a last chance of reproduction. The results of multiple invasions of brood cells are horrible to behold. Although only a small number of crippled bees were found on the floor during the experiment, the true figure may be much higher. Dead bees, young or old, which were lying on top of the screen would be carried out and dropped some distance from the hive.

When I found that the daily count had not diminished on Day 19, another strip of Folbex VA was administered. The two counts, two and eight hours after fumigation, made me gasp again and take more wine with my water. I decided to change tactics and continue the fight against the mite with strips of Apistan (Zoecon), the approved chemical control in France and USA. Apistan are strips of plastic material, soaked with the chemical fluvalinate, and this acts as a slowly evaporating 'Nano-fumigant' over several weeks without harming the bees (on the lines of the Vapona strips (but never be tempted to use that insecticide; Vapona is dangerous to bees and kills your colony!).

Apistan strips are available from the local agents for equipment, and I had hoped that the price would not break the bank. Well it nearly did, the sealed pack of 12 strips cost me 130 Francs (roughly £13) and two such strips are required for each colony - twice a year. This amounts to £5.20 per stock and is an expensive way of saving a colony. For the professional beekeeper the expense might well break the bank or break his bank manager's heart and good will. But what is the alternative? The death of all stocks after a few years.

Well, the Apistan strips did the job (see **Table 2**). Thirteen, fourteen days after placing the strips into the hive, the numbers of varroa found on the floor crashed to insignificant numbers, although the decline due to the effective Folbex VA could be noticed clearly before then. Very few bees were found dead and that number could be due to natural causes. But when the daily count (with Apistan) is dropping from an initial average of 450 mites per day to 250 (13 days after Folbex) and, finally, to less than a half dozen varroa per day, then it shows that the battle is being won. From now on I will make a count every ten days and will report again in spring. Meanwhile I will have to do more gardening instead.

Perizin is not an approved formulation in Britain, and Folbex VA is also phased out of production now. Other less harmful and probably more effective drugs may be on their way, but it appears that Apistan is hard to beat. Here in France the beekeepers prefer it to getting their lungs full of dusts and other poisons. The chemical fluvalinate is available under the name of "Klartan" and is used as an agricultural spray. Beekeepers soak strips of cloth with diluted Klartan (2.5%) and let them dry. Other beekeepers have sprayed a watery solution (0.02 - 0.05%) over the bees in August/September. These home-made control methods are not legally approved and, officially, the drug may only be used in the form of the Apistan strip. But the gendarmes are far away and not interested in what beekeepers do. When they call, they get a swig of the local Vin rouge.

Whichever drug or which chemical firm wins through, the figures of **Table 1** show that the beekeepers' fight against varroa will be a constant struggle without let-up unless some good medicine is available. Leave the colony alone for a year, use the wrong material - or the right drug at the wrong time, and a strong winter could leave you without a colony or the professional beekeeper without an income. To me it is five pounds well spent when one considers the alternative: the death of the colony valued between 50 - 100 pounds. I am certain that any colony with the number of varroa mites which I found in my colony over the 5 weeks could not survive an English winter - nor a Scottish one. My bees, on the other hand, will now rear another, healthy generation of winter bees and I am confident they will survive the short 'winter'.

Alright, I have been a varroa scaremonger. For years I have been in the vanguard of the struggle to get everybody, amateurs, professionals, scientists and administrators alike, to be aware of the problem to come. We must thank our lucky stars that, so far, varroa has not reached the British shores. When the mite is discovered, it will be too late for recriminations and for punitive 'burn-them-alive' policies. When the first colony collapses, it will be too late. When the first crippled baby bee is seen on the flight board, it will be too late. It will be too late when the first paper-inlay displays the first varroa mite in Britain.

After that it will be an annual or bi-annual fight to keep the mite from doing damage to bees and brood. It will be a constant struggle to keep the mite population within tolerable levels while keeping all hive products free from chemicals. Professional beekeepers will need a simple medication which can be applied in a jiffy, - yet can be relied upon. They cannot afford the cost of special floors with paper inlays under every hive, nor the cost of checking for the varroa mites. They cannot afford the labour costs of constant, costly and complicated medication, fumigation or dustings.

In any country where a winter can become a killer of many healthy colonies, the additional burden put on the health of a colony by the parasitic mite will be too much to bear. As things stand at the moment, the only remedy to avoid total loss is to administer a good and harmless drug on a blanket basis every year for all years to come. May that day never come your way!

P.S. There is a sequel to the story and many things have changed. For a start, Britain has discovered the mite and, as predicted, the mite had spread far and wide before the first 'outbreak' was discovered. Also, as predicted, varroa had been in Britain for years, and no paper inlay, even from the infected apiary, had revealed the presence of

The New Varroa Handbook

the mite.

Eradication is not possible, all we can do is to slow down the parasite's advance through mainland Britain and its spread to any isles.

Here in France I continued to count varroa mites on a ten day timetable and found, on average, less than one mite per day during the winter months. Pale mites were always among the dead varroa, proving that bees in this part of the world hardly stop rearing brood. The number of mites actually decreased and, by March, many a period went by without a mite on the floor.

The colony, on the other hand, was working hard throughout that time. By December the gorse (whin) produced much pollen, by mid February the mimosas helped along. I also removed frames of honey and replaced them one by one with foundation. The colony was strong enough to take one full sheet (Dadant!) 10th February, and another one was given February 22nd. A little sugar syrup was kept available on the door step to encourage brooding and wax making. A third frame with foundation was added March 6th, the honey being fed back to the bees. The colony had been getting stronger every day from Christmas onwards. Another point, the bees were also less 'stroppy'.

When a Maud queen arrived at the end of March, the decision was taken to add her to a 'shook-swarm' of very young bees only. A plastic hive had been ready with foundation for some time, and this recieved the shaken bees from 8 frames of brood. Seeing a queen cell on an outside frame, I also killed the 'French Madam' at the same time. The Maud queen was released by the young bees from her cage after the old bees had returned to the Dadant hive.

This was taken away into the Corbieres, were it was split into two nuclei, one was given to another Brit who is a winegrower there, the other became my 'out-apiary'.

The bees in their plastic hive drew out six frames in a hurry, and the queen had filled four of them out in three weeks. Measurements showed that she had laid them out at a rate of 1000 eggs per day. After 4 weeks had gone by, the new bees began to take over the wax-drawing duties and the work proceeded far faster than before.

Today, end of May and 8 weeks after the introduction of the queen, the bees (nearly all Maud-strain by now) are brooding on 9 frames Langstroth and are working in their first super. The two nuclei have mated queens and 6 frames of emerging brood and are also making use of their supers, although a 'June gap' has slowed things down just now. By the way; none of the colonies have a single cell of AFB.

Who says there is no life after varroa?

With the right treatment, good bees and good, yes, BETTER beekeeping, it is possible to carry on keeping bees and getting good honey crops.

Bernhard Mobus

TABLE 1

2.1

VARROA CHECK 1991

		PERIZIN and FOLBEX VA		Total	Adult (drk)	Young (pl/br)
Day 0	Drug	10/09/91	PERIZIN, 1ml in 50 ml H_2O			
			Collection 5hrs after medication	69	58	11
Day 1*		11/09/91	3 Collections	275	164	111
Day 2		12/09/91		79	18	61
Day 3		13/09/91		84	12	72
Day 4		14/09/91	Crippled bees 1	87	11	76
Day 5		15/09/91		43	9	34
Day 6		16/09/91	Collection before medication	81	15	66
	Drug		PERIZIN, 1ml in 50 ml H_2O			
			Collection 5hrs after medication	127	54	73
Day 7*		17/09/91		140	42	71
Day 8		18/09/91		85	19	68
Day 9		19/09/91		70	14	56
Day 10		20/09/91		73	16	56
Day 11		21/09/91		119	30	69
Day 12		22/09/91	Crippled bees 2	88	28	60
Day 13		23/09/91	" " 3	68	18	50
Day 14		24/09/91	" " 1	61	17	44
Day 15		25/09/91	Collection before fumigation	114	29	85
	Drug		2 strips FOLBEX VA			
			3 Collections after fumigation	615	Mainly dark	
Day 16*		26/09/91		121	49	72
Day 17		27/09/91		222	66	156
Day 18		28/09/91	Crippled bees 4	176	51	125
Day 19		29/09/91	Collection before fumigation	167	42	125
	Drug		1 strip FOLBEX VA			
			2 Collections after fumigation	581	Mainly dark	
Day 20*		30/09/91		158	45	113

TOTAL KILL: 3703

* Less than 24 hrs after previous collection

The New Varroa Handbook

2.2 TABLE 2

VARROA CHECK, 1991

			APISTAN (Zoecon)				Total varroa	Young (pl/br)	Dead bees
Day 20	Drug	30/09/91	2 strips APISTAN, applied over frames,				7pm		
Day 21		01/10/91					447	+	5
Day 22		02/10/91					483	+	
Day 23		03/10/91	Crippled bees		1 + *		357	+	2
Day 24		04/10/91			*		390	+	4
Day 25		05/10/91					430	+	1
Day 26		06/10/91			*		370	+	4
Day 27		07/10/91	Crippled bees		3 + *		533	+	1
Day 28		08/10/91	(Should reflect 1st FOLBEX treatment)				289	+	4
Day 29		09/10/91					196	+	
Day 30		10/10/91			*		276	+	
Day 31		11/10/91			1 + *		269	+	
Day 32		12/10/91	(Should reflect 2nd FOLBEX treatment)				203	+	1
Day 33		13/10/91			1		102	+	
Day 34		14/10/91	(Should reflect APISTAN treatment)				68	+	
Day 35		15/10/91	I	I	I	1	5	+	
Day 36		16/10/91	I	I	I		10	+	
Day 37		17/10/91	I	I	I		1	-	
Day 38		18/10/91	I	I	I		6	+	
Day 39		19/10/91	I	I	I		6	+	
Day 40		20/10/91	V	V	V		3	+	

Total varroa count, APISTAN		4444
Prev. varroa kill, PERIZIN & FOLBEX VA		3703
Total varroa kill, all drugs		8147

* Floor debris revealed parts of immature bees, probably crippled by varroa

LEARNING TO LIVE WITH VARROA IN THE UK -
Clive de Bruyn

3.1 INTRODUCTION

Varroasis is finally here in the British Islands. It is undoubtedly here to stay, eventually it will become just another endemic honeybee parasite. It is not likely to wreak the havoc alarmist commentators of the sensationalist media like to predict. On the other hand the problem cannot be considered trivial. Varroasis is a serious threat to beekeeping. At this time of writing it is still a notifiable disease by law under the Bee Disease Control Order 1982.

Beekeeping is still continuing in all other countries where the varroa mite has arrived. Once the mite is discovered in an area it will spread unremittingly throughout the land mass until every colony is infested. We have the advantage of being in a position to benefit from the knowledge and expertise gained from other beekeepers who have been through this experience already.

Despite recommendations promulgated widely in the last 10 years to search for the mite, when it was discovered the numbers of mites indicated that varroa had been present in the colonies for several years. If all beekeepers had been more diligent in searching for the mite it could have been possible to eradicate the invasion before the mites had become widely spread over an area.

Many beekeepers are still not sufficiently clued up. They need to be searching every colony regularly so that they are aware if their colonies are infested or not. When mites are found, as they will inevitably be, the beekeeper should have prepared a strategy on what to do.

Searching for mites and monitoring levels of infestation will now become part of our routine management every bit as important as swarm control, removing the crop and feeding. The well known diseases will also not disappear we cannot afford to neglect foul brood and the minor brood diseases. Stress is a precursor of many ailments. Heavy levels of varroa can kill a colony, low levels will act as a stress on the colony.

It is not every generation of beekeepers that are privileged to play a part in such exciting times. The arrival of a new parasite onto a fresh host is an interesting study. The rate at which the mite spreads, its effect on beekeeping, honey production and pollination we have still to find out. I have often wondered what actually happened in 1904 to 1914 when the "Isle of Wight Disease" raged. I do hope that more careful records will be kept of the impact that varroa makes in the next few years.

In the last 15 years I have been able to

inspect colonies with varroa mites in Brazil, China, France, Germany, Greece, Japan, Poland and Thailand. I have been fortunate enough to discuss the problem with beekeepers from all over the world at international beekeeping conferences. Some of the immense range of literature on the topic has also been studied. Despite this I still feel I will only begin to know something about varroa and its effect on the honeybee when I have managed infested colonies of my own for a decade.

Despite the fact that the mite has been known about for so long there is much confusion and contradictory evidence about its biology, its damage to bees and the best way to control it.

It might be considered best to wait until I was more sure of the facts before going to print. But, if facts are all that matter then no one would ever write anything about bees. I remember thinking how simple keeping bees was in my third year of taking up the hobby. Since then with every passing year I feel that I know less than I did the previous year.

No doubt my opinions about varroasis will change as we learn more about the mite in this country. Before this occurs I think it would be useful to record the present state of affairs. Future generations might well be interested in these early days.

It is also necessary to discuss what is known about varroa and describe some of the controls which others have used.

Beekeepers are fiercely independent and it is not likely that any one strategy will be universally adopted. Decisions on what to do will also be dictated by the number of colonies you manage, your experience, your moral and ethical disposition and to some extent what all about you are doing. In the following pages I will try to present some of the options which are available.

The Varroa Experience

In talking with other beekeepers about what happened when the mite first arrived in their country some common factors have emerged. When the first find is made there is generally a period of over-reaction when beekeepers are angry, frightened and confused. At this time fingers are pointed and doom prophesied. The media, always interested in sensationalism, does its best to whip up the situation.

This scenario is repeated as the mite spreads into each new area until the entire country is infested. A few beekeepers give up when varroa is found in their hives. Others turn to control measures proven successful elsewhere. Some are too rigid to adapt and they just sit back and wait.

When it is found that colonies can survive if managed properly everyone relaxes. A few years pass and everyone is convinced that things were not as bad as had been predicted. The expected apocalypse never comes and beekeepers revert to their bad old ways. In this time of complacency they stop monitoring or treating their bees.

The New Varroa Handbook

Then when colonies which are not looked after start to die another group of beekeepers give up. Some let alone beekeepers lose their bees and start up again without changing their management. How many times they will restock without learning is not certain.

Eventually the country is left with a hard core of practising beekeepers. Will you be one of them? Beekeeping is still continuing in all other countries where the mite has spread to. Colonies can survive and be productive provided that beekeepers are aware of the infestation and are willing and able to adapt their management accordingly. We have the advantage of being in a position to benefit from the knowledge and expertise gained from others who have been through this initiation already.

Apart from looking after our bees it should be in all our interests to reassure the general public that there is no danger to human beings from the mite. It is not a disease that can contaminate honey. Beekeepers will eventually learn to cope. It would also be sensible to spare a thought for your fellow beekeepers. Let alone beekeepers, Association types, commercial or professional beekeepers and anyone who dabbles in bees we are all in this together. Helping your fellow beekeeper is not altruism but self interest.

Clive de Bruyn

NATURAL HISTORY   3.2

Varroa jacobsoni is native to Asia where it infests the Indian Honeybee, *Apis cerana*. This species and varroa have had a long association during which they appear to have developed a tolerable relationship. Geographical distribution of the Indian honeybee and varroa is from China to Indonesia and from Afghanistan to Japan. These are all areas where the Western honeybee, *Apis mellifera* is not naturally present.

An organism that lives as an ectoparasite on another species nourishes itself at the expense of the host without rapidly destroying it, although often inflicting some degree of injury **Ref 1**

Varroa presents a problems to the mellifera species as neither the mite nor the host have had sufficient time to adapt. Varroa is not a good parasite on the Western Honeybees as it appears to be responsible for eventual colony demise. Good parasites should get on well with their hosts causing minimal damage. This is the case in *A. cerana*. Where the host only suffers slight damage the parasite's own future is made secure.

In the Indian Honeybee varroa appears to breed exclusively in the drone brood. Although the female varroa enters worker and drone cells she only lays eggs in the latter. **Ref 2, 3**

The adult female varroa is a relatively large mite with four pairs of short segmented legs barely visible from above. Using a magnifying glass the legs can be seen if the specimen is turned over onto its back. It is usually described as reddish brown but I have noticed they can vary from brown through to orange. It is flat smooth bodied animal resembling a crab, which I have no problem seeing with the naked eye. It is shorter head to tail than from side to side. Its mouth parts are adapted for piercing and sucking.

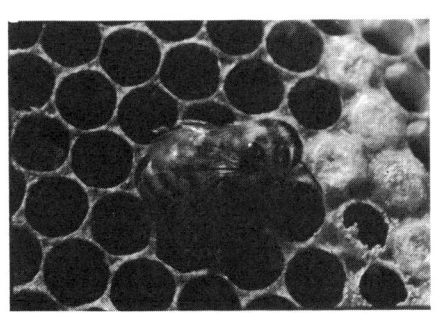

Fig 1 *Worker with mite on its back*

The female mite has to feed on the haemolymph of the brood before it can lay eggs. The rate of reproduction is influenced by the endocrine system of the honeybee. The juvenile hormone (JH) necessary to trigger egg laying occurs in greater concentrations in drone larvae than worker larvae. This is particularly noticible in the case of *A. cerana*. where the workers have a

The New Varroa Handbook

comparitively low JH. Low concentrations of JH inhibit mite reproduction. JH varies between species sex and season. JH is higher in *A. mellifera* than *A. crerana* and in summer bees compared to winter bees. **Ref 4**

In the first 60hrs the drone larvae of *A. cerana* and *A. mellifera* contain more than 5ng/mlJH in their haemolymph. Worker larvae of *A. mellifera* contain 3-7 ng/ml and those of *A. cerana* contain only 1ng/ml. This level of JH is apparently insufficient to induce oviposition in the mite.

Estimated life of varroa in the UK

In Summer	2-3 months
In Winter	5-8 months
Without bees and brood	5 days
In sealed combs of brood at 20°C	30 days

Ref 5, 6

Because of the unique biology of the parasite a detailed study of its reproductive behaviour has not been easy. It is extremely difficult to study feeding, mating and reproduction behaviour when the organism is sealed in an opaque cell sequestered within a honeybee colony. The life cycle can be divided into different phases. **Ref 7**

Phase I
The phoretic period when the female mite lives on adult bees.

When the brood cell is uncapped fully developed female mites, fertilised or not, can emerge from the cell on their own or attached to a bee. The mite can move quite fast. This can be quite frightening when seen magnified under a microscope. In 1985 I was able to visit a W. German research institute with students from Cardiff University. We were allowed to take some sealed drone brood from an infested colony to look at in the laboratory. I well remember the fright I had when I first saw the mites come tumbling out of a drone cell which I had just uncapped. The sight of the mites gave me many an itchy feeling for some time afterwards.

The female mite after emerging search for an adult worker or drone host. They burrow between the overlapping segments under the bee's abdomen (sternal plates). With her mouth parts (Chelicerae) the mite pierces the intersegmental membrane between the honeybee's cuticle plates so that she can feed on the bee's haemolymph. Mites are dispersed through the colony on adult bees in the colony.

Choosing a house bee as a host is more advantageous for *Varroa jacobsoni* because these bees remain in the brood area where the varroa reproduces. It would appear that there is a difference in Nasonov pheromone production as bees age. Female varroa mites are repelled by Nasonov pheromone which increases as bees grow older. The mite is therefore capable of recognising younger bees **Ref 8.**

She feeds for several days on the adult bee. (4-13 days). Much more information is needed about the duration of this phase and of the number of mites that are able to re-infect larvae more than once. Such factors would have an influence on the population dynamics of varroa. It could be that these variables change with the seasons, the level of infestation and the type of bee.

Phase 2
Penetration into larval food and inactivation.

After the phoretic period the fertile mite, with eggs already in her ovaries, (gravid) leaves the adult bee to enter a brood cell just before it is due to be capped. (1-2 days before sealing?). Young varroa females enter cells after 5-15 days of life on adult bees. Older reproductive varroa mothers can re-enter brood cells quickly for another cycle. Given a choice the mites prefer drone brood to worker.

Female varroa mites rarely enter queen cells and deposit eggs. Under no circumstances do varroa mites have time to develop fully or reproduce in queen cells. More than one mite can enter a single cell but this generally only occurs when the infestation is heavy. In excess of 20 mites have been found in a single drone cell **Ref 9.**

The varroa mite hides in the brood food where it enters a cataleptic phase.

Phase 3
Reactivation feeding and egg laying.

Once the cell is sealed the bee larva consumes the remaining larval food before it changes into a pupa. Normally the varroa mite emerges from her cataleptic state when all the brood food is consumed. The mite now moves onto the prepupa where it starts to feed. It uses its piercing mouth parts to pierce the pupal cuticle so that it can feed on the haemolymph. During this period the first egg grows to maturity. Not all females that enter brood cells are able to reproduce. In experiments 50% of the mites failed to reproduce at all in Africanized colonies. Only 15% failed to reproduce in Carniolan colonies. 20% of mites fail to reproduce in Germany **Ref 10.**

To show how confusing the situation is there has even been work in Russia which reports that the varroa does not lay eggs but gives birth to larvae. **Ref 11**

The factors effecting oviposition still need further study. According to the data collected so far. Fertile female mites that move into brood cells lay their first eggs about 60 to 64 hours after the cell is capped. A significant proportion of first eggs laid are male, second and subsequent eggs are predominantly female. **Ref 12**

The first female mite completes her development in 170 to 200 hours from when the egg was laid. Males develop in a shorter period of time. Yet another 24 hours are needed before the outside surface (cuticle) of the mite is hardened and tanned. Only after this occurs can the mites mate and move out of the cell. Whilst the first female mite that develops might have the time required, her sisters

The New Varroa Handbook

that result from eggs laid later on may not. Feeding and egg laying probably stops when the pupal eyes become darkened. **Ref 13**

Mite Reproduction
Male mites
5.5 to 5.7 days egg to adult
Female mites
6.2 to 7.5 days egg to adult

Phase 4
Laying of further eggs and embryo development

The average number of eggs laid is just over 4 per mite irrespective of the time of year or the nutrition of the colony. 50% of mites laid 5 eggs. The mature fertilised female mite has the potential of laying at least five eggs in a worker cell and seven in a drone cell. It does seem strange that the varroa should expend so much energy and resources in producing eggs that will never mature into adult mites?

When more than one varroa mite enters a cell not all of them necessarily reproduce. In the case of multiple parasitation the average number of eggs/mite decreased with the degree of infestation.

1 mite   Average 4/mite
2   "   3.5

With significant decrease of average number of eggs as numbers of mites rise. This decrease could arise from some kind of competition be tween the mites in the cell. We do not know who is responsible for the eggs laid. Are all the mites laying fewer eggs or are some mites not laying at all? **Ref 14**

The number of female mites that are able to complete their development and mate before the parasitized worker can emerge determine the build up of the mite population. **Ref 15**

The number of progeny that a single mite can produce varies according to the time the cell is capped. In the African races the postcapping stage is less and consequently the mite is not able to reproduce as fast. **Ref 16**

Brood cell post capping time	
A.m. carnica	12. days
A.m. scutilata	11.2 days
A.m. capensis	9.6 days

The mite development stages consist of egg (larvae), protonymph, deutonymph and adult. It is now thought that the larval stage occurs within the egg and is of short duration. The first female varroa mite which hatches matures in the cell. Mature males are able to mate incestuously with their sisters (1 invasion per cell) or other mature females (more than one invading mite per cell) The male uses his chelicerae to introduce his spermatocytes into the female. The male and the immature females die soon after the brood cell is uncapped.

Adult impregnated female mites have -been recorded as able to reproduce 6-7 cycles of brood. i.e. 30 offspring from 1 mother in her lifetime?

The above phases proceed

chronologically in summer one after the other. In N. European temperate regions where there is a break in brood rearing in winter the adult mite is able to survive on adult bees without brood for many months. Without bees or brood the mites life expectancy is only a matter of days.

Ref 1   Cheng. I. C. (1967) The Biology of Animal Parasites: Saunders, Philadelphia pp 3-28.
Ref 2   Koeniger, N.; Koeniger, G.; Wuayagunasekara, N.H.P. (1981) Beobachtngen uber die Anpassung von *Varroa jacobsoni* an Ihren naturlichen Wirt *Apis cerana* in Sri Lanka Apidologie 12: 37-40
Ref 3   Tewarson. N.,C., (1987). Use of host haemolymph proteins. seasonal reproduction and a hypothesis on nutritional imprint ing on the honeybee mite *Varroa jacobsoni* on *Apis mellifera* and *Apis cerana*. in: Chemistry and Biology of social insects. Proc of 10th International Congress IUSSI Muenchen, Fed Rep Germany 688-689.
Ref 4   Hanel, H., (1983) Effects of JH III on the reproduction of *Varroa jacobsoni*. Apidologie 14, 137-142
Ref 5   Shabanov. M., Nedyalkov, S.T. Tashkov A.L. (1978) Varroatosis- A Dangerous Parasitic Disease on Bees: American Bee Journal June 1978 402-403
Ref 6   Ritter, W. (1981) Varroa Disease of the Honey Bee *Apis mellifera*. Bee World 62 (4): 141-153
Ref 7   Ramirez W.B. & Gard W. (1986) Development phases in the life cycle of *Varroa jacobsoni*, an ectoparasiote mite on Honeybees. Bee World 67(3): 92-97.
Ref 8   Hoppe. H. & Ritter W., (1988) The influence of the nasonov pheromone on the regulation of house bees and foragers by *Varroa jacobsoni*. Apidologie 19 (2): 165-172
Ref 9   De Jong,D; Morse, R.A.; Etckort, C.GF.; (1982) Mite pests of honeybees. A. Rev Ent. 27: 229-252.
Ref 10   Kulincrevic. J.M., Rindered. T.E., Urosevic. D.J. (1988) Seasonally and colony variation of reproducing and non-reproducing *Varroa jacobsoni* females in western honey bee (*Apis mellifera*) worker brood. Apidologie 20 (2): 173-180
Ref 11   Akimov, I.,A., et al (1990) Russian magazine Zoological Report
Ref 12   Rehm. S.,M., Ritter. W.,(1989) Sequence of the sexes in the offspring of *Varroa jacobsoni* and the resulting consequences for the calculation of the developmental period Apidologie 20: 339-343.

Ref 13 Grobov, O.F. (1977) Varroasis in Bees Pp. 46-70 from Varroasis, a honeybee disease, Bucharest, Rumania, Apimondia.
Infantidis, M.D. (1982) The onto genesis of the mite *Varroa jacobsoni* under natural breeding conditions and factors influencing its reproduction and population growth. Univ. Thessaloniki, Thessaloniki, Greece: Infantidis, M.D. (1983) Onto genesis of the mite *Varroa jacobsoni* in worker and drone honeybee cells. J.apic Res. 22: 200-206.
Ref 14 Gonzalis et al (990) Apimondia personal communication
Ref 15 Infantidis, M.D. (1983) Onto genesis of the mite *Varroa Jacobsoni* in worker and drone bee brood cells. J. Apic. Res. 22 (3):200-206
Ref 16 Moritz, R.A.; Hanel, H. (1984) Restricted development of the parasitic mite *Varroa jacobsoni* in the Cape honeybee. Z. a ngew. Ent. 97: 91-95

Clive de Bruyn

WHAT EFFECT WILL VARROA HAVE? 3.3

When the nymphal mites feed on the honeybee pupae they damage their hosts. The feeding activity of a large number of adult and nymph mites can produce bees that are smaller, and deformed, with a reduced life span **Ref 1, 2** The number of mites in a colony build up steadily over a period of three to five years at which time colonies succumb. The exact mechanism how the colony crashes and the bees die is not precisely understood. Secondary infections and pathogens have been implicated. **Ref 3**

Varroa infestations can weaken adult bees leaving them susceptible to concurrent viral infections. The brood is also damaged resulting in deformed bees. Drones heavily parasitised by varroa will not fly they also have a lower sperm count. Because the drones are so susceptible to varroa it is probable that this mechanism will assist in developing resistance to varroasis.

When the majority of the nurse bees are infested as pupa they are not capable of feeding the brood properly. Endemic diseases which are exacerbated by stress may then gain a hold. Two brood diseases in this category are European Foul Brood and Chalk Brood, both of which have been associated with varroasis. The complex interactions in the colony between the bees and various pathogens are difficult to understand. Chalk brood is a significant problem in places with warmer climates such as Greece and Portugal where varroa is widespread. Serious losses have been recorded in these countries. Colony losses in Germany have been shown to be caused by viruses spread by varroa mites. **Ref 3**

The reported damage that varroa causes varies widely from country to country. Bees, beekeeping practices, climate and geography can all influence what effect varroasis will have. In addition to this must be added the complications brought about by beekeepers treating their colonies to control the varroa mites. It is also difficult to obtain reliable data which can be separated from hearsay. We will only know what effect varroa will have on our colonies if we bother to monitor mite levels and keep accurate records ourselves.

As a rough estimate the number of mites in a colony doubles in every two cycles of brood. Added to this is the number of mites still alive after over wintering. The more prolific the bee in your hive the greater will be the amount of brood present in the active season (April to September). The sooner the rearing of drones begins the larger will be the area of drone brood for mites to breed in. The more brood there is the quicker the mite numbers will reach damaging proportions.

If special spring stimulation is given then the colony will build up faster leading to a faster population growth

The New Varroa Handbook

of the parasite. The point of collapse may in such instances take place after two years rather than four. Beekeepers who practise spring stimulation with pollen supplement and judicious brood spreading will have to think carefully what effect this will have on the mite population.

Good weather combined with mild winters can be favourable to mite development. Good honey flows can restrict the brood area confining the mites reproductive opportunities. This can mean that all the mites were concentrated in a proportionately small area of brood resulting in a high infestation of individual pupae and the subsequent emergence of a significant proportion of damaged and crippled adults.

Paralysis

Acute paralysis virus (APV) has been found closely associated with an infestation of varroasis. The role of viruses in honeybee pathology is not fully understood but is likely to be more important than people think. Few countries posess the technical sources or expertise to monitor bees for virus diseases. Rothamsted Experimental Station has been working on virus diseases of bees for many years, firstly under Leslie Bailey and now Brenda Ball. **Ref 4**

Research in collaboration with European colleagues has established that APV is a primary cause of adult bee and brood mortality in A. mellifera colonies severely infested with V. jacobsoni. To date APV has never yet caused mortality in Britain. Field and laboratory studies indicate that APV can multiply in varroa mites who also transmit the infection to their hosts. Now that varroa is present here future work by Rothamsted can be carried out to determine a better understanding of the dynamics of the virus infection and its interaction with varroa infestations.

Ref 1   Arnold, G. (1990) Current and recent research on Varroa in Europe. American Bee Journal 130: 257-261
Ref 2   De Jong, D.; De Jong,P. H.; (1983) Longevity of Africanized honey bees infested by Varroa jacobsoni. J. econ. Ent. 76: 7 66-768. De Jong, D.; De Jong, P. H.; Goncalves,L.S.; (1982) Weight loss and other damage to developing worker honeybees from infestation with Varroa jacobsoni. J. apic Res. 21: 165-167. De Jong, D.; Goncalves,L.S.; Morse R.A. (1984) Dependence on climate of the virulence of Varroa jacobsoni: Bee Wld 65: 1 17-121.
Ref 3   Ball, B. V. (1985) Acute paralysis virus isolated from honeybee colonies infested with Varroa jacobsoni. Journal of Apicultural Res. 24: 115-119
Ref 4   Ball, B. V. (1990) The Impact of Secondary Infections in Honey Bee Colonies with the parasitic mite Varroa jacobsoni Rothamsted Experimental Station, report.

Clive de Bruyn

THE DETECTION OF VARROASIS 3.4

At present only a pitifully small number of beekeepers in the United Kingdom have been looking for the mite in their colonies. No one can say with any certainty that varroa is not present, anywhere in the British Islands. I would hope that in the future all responsible beekeepers will accept "Detection and monitoring of colonies for Varrroasis as a normal aspect of colony management".

Treatments used in other countries will not necessarily be suited to our bees, our climate, our hives or our system of management. It will therefore be important that beekeepers become efficient in finding mites and assessing the number present in a colony. There may be great differences in infestation levels. We know already that the rate at which varroa mites build up in a colony can vary. This suggests that some bees may be more susceptible to varroa infestation than others. In this

Table 1

Voluntary Samples sent to the National Beekeeping Unit for Varroa search (England and Wales)

Year	Total Number of Beekeepers	Number of Beekeepers who submitted samples	Number of samples
1988	31,036	1007	1302
1989	30,352	752	920
1990	30,967	541	658
1991	28,313	924	1044
1992		Not issued yet.	

Beekeepers will need to know when the parasite becomes established in their area. Once present in the apiary it will be important in the early years to make regular examinations of all colonies to assess the number of mites present and evaluate the state of infestation. Without such monitoring the success of any treatment can only be speculative.

early period the more information that can be obtained about the population dynamics the better equipped we will be to select a suitable treatment and make a choice about when to act. Some initial work may also be possible in detecting colonies which show some "resistance" to the parasite.

We do not know how fast varroa will

The New Varroa Handbook

move in the UK. It could be interesting to monitor its spread. Whether the mites' presence has been confirmed in a district or not I consider it will be irresponsible if beekeepers did not search for it in their colonies from now on. One cannot rely on other beekeepers or the Ministry. This is an individual problem for everyone who keeps bees. Know where your enemy is!

Routine colony inspections will not reveal the presence of varroa until it is too late. Beekeepers will need to assess the different detection techniques and refine the various methods to suit their own circumstances. If possible they should choose those that do least harm to the bees or the produce of the hive and yet fit in with their normal system of management without incurring undue expense or labour. It is important that varroa is detected early before noticeable damage to the colony occurs.

Varroa can be found:- on adult bees, on the brood, in comb or in hive debris. The number and location of mites varies with the season and the time of the year. In our climate one would expect that the number of mites would be lowest in spring steadily rising through the summer to reach a peak in the autumn. During spring and summer most mites would be found in the brood especially the drone brood. In late autumn and winter mites will be found predominantly on adult bees. Positive sighting of the mite confirms its presence. The fact that no mites are seen does not indicate that they are not present. An element of luck is needed!

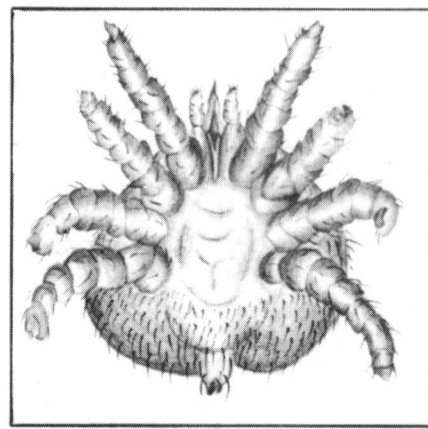

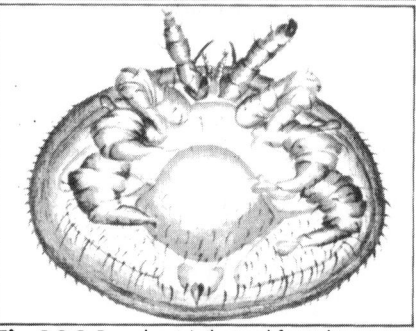

Fig 2&3 *Braula*. Male and female varroa

Before you start looking for the mite you need to know what the varroa female looks like. As mites go we are fortunate that varroa is a relatively large mite. Anyone who can see the eggs laid by the queen in the colony should have no problem in seeing the female varroa. It will also be important to distinguish between the bee louse *Braula coeca* and the mite *V. jacobsoni*. They are of similar size, visible to the naked eye, but shaped differently. Braula is an insect with six legs, three on each side. It is found only on adult bees not in brood cells. Varroa is an arachnid with eight forward extending legs, it can be found on adult bees and in the brood. Personally I do not consider that counting legs is a sensible way to distinguish between these two creatures. The main difference between braula and varroa is in the shape. Many of the methods used to detect Varroa will also reveal braula. If you do find braula using any of the techniques given then you will have some guide that the technique is valid:- at least for *Braula coeca*.

There are literally 100s of methods available some of which are more suited to research institutes than practising beekeepers. Fortunately they are mostly variations on a few themes. A few of the more common techniques currently used to detect the mite will now be described.

Visual Examination of Bees

Fig 4 *Examining individual bees for mites*

Examination of adult bees is not a very accurate or reliable method. The number of mites on adult bees will vary throughout the season. I remember the time in 1985 when a party of students from Cardiff University and I visited a research institute in Germany. We opened up a colony, which was supposedly moderately infested, and picked up individual worker bees by hand. We inspected them all over for mites and had to search several tens of adult worker bees from the colony before we spotted our first varroa mite. **Fig 4**. When I have seen varroa moving about on bees they have not been difficult to see. This is not always the case when they attach themselves between the abdominal segments.

Fig 5

Examination of a Sample of Bees

A sample of adult bees can be collected from a brood frame containing newly emerged workers. First locate the queen and make sure she is left safe in the hive. 300 to 500 can be brushed off the frame or shaken through a funnel into a container (honey jar). The frame can now be returned to the brood nest and the hive closed up. The bees in the jar are shaken with an agent which will dislodge the mite. Alcohol, diesel fuel, petrol, hot water, or detergent solution have all been found to work with differing degrees of success.

The sample can be scanned through the sides of the jar. A better method of inspection is to pass the material through a screen with openings approximately 4mm square (small enough to retain bees but large enough to allow mites to pass through). It is always a good idea to wash the bodies of the bees and the screen with fluid to ensure that no mites are left adhering to the bees or container. The fluid (+ mites) passing through the screen is poured onto two thickness of white cheese-cloth. The liquid can thus be collected for future use. Any varroa or braula present should show up on the cloth.

A popular method used in the USA is the so called "Ether roll". The procedure has some similarities with the previous technique. The major difference is the use of an aerosol containing ether instead of a fluid. An aerosol formulation is available in England as an aid to starting cars in cold weather. (Check the label for ingredients) The ether is sprayed into the jar of bees whilst the lid is slightly opened to prevent mass escape. One or two short bursts of the aerosol just under the lid is usually sufficient.

Fig 6

The lid is firmly screwed on and the jar is slowly rotated for about 30 seconds. Both the bees and any mites are anaesthetized. This dislodges the varroa from the bees causing them to stick to the walls of the container. To complete the process the sample can be sieved as before and deposited on a white surface where the mites will show up.

These methods are rapid, relatively cheap and simple. The answer is provided straight away. As only a small sample of bees is used there will be a high probability that low levels of varroa infestation will go undetected. The method involves killing bees directly and is not recommended for the squeamish. On the other hand it does not involve introducing chemicals into the hive.

Examination of the Brood

A method of examining brood for the presence of the mite is to use a capping scratcher as used in uncapping. Drone brood is supposedly more likely to be infested than worker brood though I have not always found this to be the case**. The sealed brood is uncapped when the pupa is at the dark eye stage. The mature varroa, if present, will have coloured up making them visible against the white background of the pupa. The inside of the cell can also be examined for nymphs. Examination of at least 100 pupae is recommended. This is a time consuming method which is also labour intensive. It is totally unreliable for early detection and can only be used when brood is present. On the positive side, no special chemicals or equipment are necessary.

** In August of 1992 I found more mites in worker brood cells than drone brood in infested colonies on the Isle of Wight.

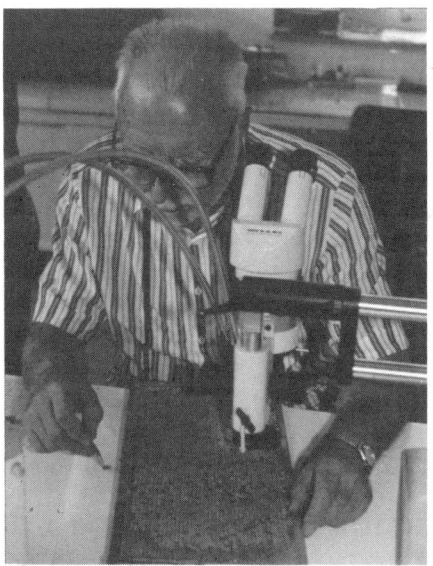

Fig 7 Searching for mites in sealed brood under the microscope

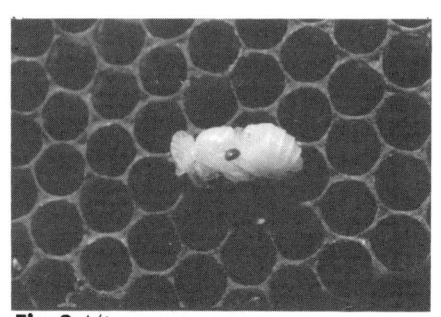

Fig 8 Mites on a pupa

Examination of Hive Floors for Dead Mites

A check for dead mites which fall off bees in the hive is a proven technique. In the spring and summer collecting the rubbish that falls onto the hive floor is not possible without a screen to prevent the bees removing the debris. A good time to collect debris and dead mites which accumulate on the floor is at the end of winter before the bees do their spring cleaning.

> Dear Clive
> Hi. My first floor sample and I muffed it. Separated the floor and the hive easy enough, bees weren't too upset. Scraped the floor debris out, away from the hive, into a shoe box. Took the floor back and rebuilt the hive all within two minutes, thought I'd done pretty good. Set hive right and turned round to take scrapings indoors and saw the shoe box upside down on the grass. The wind had caught it and upturned the contents into the lush uncut damp spring lawn! Hope botanical samples don't hinder too much and sufficient debris remains.
> See you on the next course
>
> **A Student**

In spring many beekeepers replace their hive floors at their annual "Spring Clean". During this operation it is a simple matter to brush the accumulated rubbish on the floor into a suitable receptacle. (**Fig 9**) I say simple but

The New Varroa Handbook

beware of wind and rain whilst collecting floor scrapings, they can easily be lost. In some years (1990) an early start to the season can catch beekeepers out. There may be no debris as the house cleaning bees have removed everything from the floor before the beekeeper is ready. You have been warned!

The operation can take place as early as February in some areas. It does not involve a full inspection of the colony with comb removal. The brood box is freed from the floor and stood to one side (on top of an upturned roof). The old floor is removed and replaced by a clean one. The brood box is then returned onto the clean floor in its

Fig 9 *Brushing winter debris from the floor*

original site. Once the debris has been collected the old floor can be scraped clean and is ready for use on the next hive. When all of the colonies on site have been dealt with the accumulated debris can be labelled with the apiary site and date.

To ease the work of examining the debris you can reduce the bulk of the sample by passing the material through a coarse sieve to remove the dead bee corpses, mice droppings, wood propolis and wax fragments. At this time debris can still be sent to the

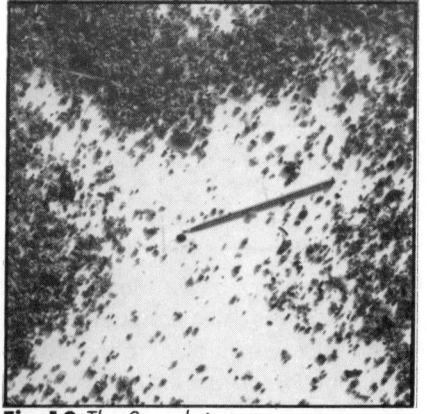

Fig 10 *The Search is not easy*

Clive de Bruyn

National Beekeeping Unit (NBU) for examination. **Ref 1**. I would expect that you could attend a course in your area to learn how to examine the debris yourself. Local associations should be able to arrange "Varroa Workshops" where beekeepers can gather and do their searches together. People who do their own screening may care to inform the NBU as to the fact as the Ministry will not know how many people are doing their own Varroa checks unless they are told.

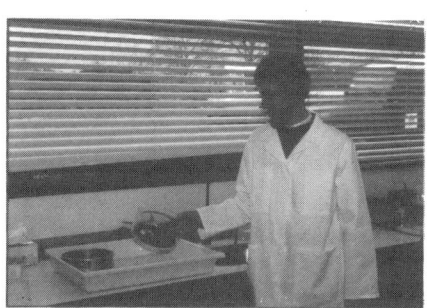

Fig 11 *Examining debris at the National Beekeeping Unit*

Fig 12 *Separation and floatation*

Material sent to the NBU is examined using a floatation technique. The debris are placed in a container with alcohol to separate the mites from the other dross. If varroa is present it will float to the top whilst the bee parts, pollen etc will sink. You can do it yourself at home, it is just like preparing a cocktail. Half fill a honey jar with a 50:50 mixture of water and methylated spirits. Add debris. Shake thoroughly and allow to stand. The debris will sink and any varroa present should float. Now that the mites are available you can check the method by putting a few mites into a "clean" sample of debris and verifying that you can find them afterwards. **Ref 2**

Use of Floor Inserts

Another simple method that can be used at any time of the year to detect mites that die and drop off bees is to insert a piece of stout card onto the hive floor for a few weeks. The card should be covered in petroleum jelly or some other material that will trap the mites. I posses a jar of horse fat from France which is supposedly 'merveilleux' for trapping varroa mites. A piece of "Fablon" sticky side up is probably a better alternative. Without something sticky to prevent the bees removing the dead mites, the method is not reliable.

In the USA there is a device called the Dewill Varroa mite detector. This is a special floorboard with a 2mm wire mesh, large enough to allow mites through but not bees, set in a frame 10mm over the base. A cardboard insert covered with a glue is supplied with the floor. On the continent there are special plastic trays and screens available for this purpose. These trays are now available in Britain and the

95

The New Varroa Handbook

appliance trade have several other options available. It should not be beyond the ingenuity of the average British beekeeper to construct such a screen and collecting tray which will work in the type of hive being used.

Fig 13 *Summer insert tray as used in Germany*

Fig 14 *Photographing debris falling down after five days*

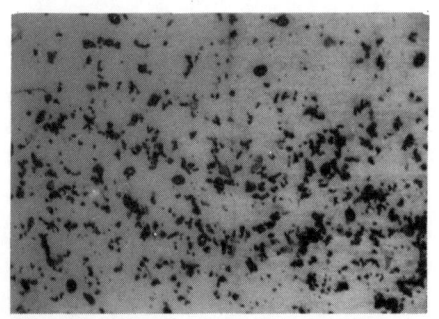

Fig 15 *What the camera saw. Spot the mites.*

Examination of inserts can be a slow and tedious procedure.

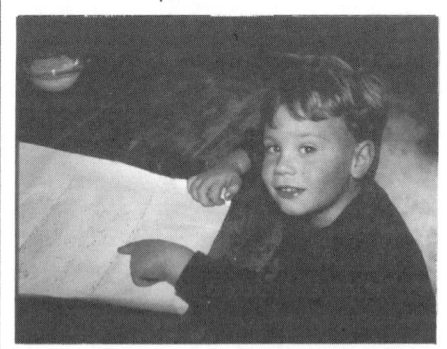

Fig 16 *Detection of low mite levels depends to some extent on skill and experience.*

Use of a Floor Insert and a "Knock Down" Acaricide

This method allows the total colony to be checked. It is one of the best techniques to detect low levels of mites, such as would be present in the initial stages of any infestation. An insert is placed on the hive floor and examined for mites after the acaricide is applied. Choosing a suitable insert and getting it into the colony and removing it afterwards presents problems to some beekeepers. Perhaps they have been following the recommendations to use wallpaper lining paper. Personally I would want something far stronger, stiffer and more resistant to the effects of damp. A stiff sheet of cardboard or plastic, preferably with a rim is much easier to handle, and it can be used repeatedly.

Many acaricides currently used for treating Varroasis have been used. As there is always a danger of contaminating the honey or wax with chemical residues it must remain with each individual as to whether or not such chemicals are introduced into the hive when the bees are collecting nectar and converting it into honey. Where this is an important consideration the safest period would be outside the major honey flows.

Originally tobacco was the only material legally available in Britain. An ADAS leaflet is available **Ref 3** setting out The NBU's method which is carried out at night when all the bees are in the hive. Tobacco is a relatively slow acting acaricide but it has a low risk of contaminating hive products. It is also easily available and reasonably cheap. Personally I see no reason to use tobacco smoke now that we have a registered acaricide which is more reliable.

In Brazil I was impressed with an acaricide smoke strip available from Yugoslavia. The "knock down" effect was most impressive. A great number of infested hives were tested in several apiaries at different times of the day when the bees were active. The bees were not shut in yet in every case mites were found on the card insert only minutes after the smoke was introduced. None of the colonies appeared any the worse after the treatment. Unfortunately due to the present situation in Yugoslavia I have not been able to ascertain what chemical the paper had been impregnated with.

Fig 17 *Inserting the paper*

Fig 18 *Igniting the strip*

Fig 19 *Inserting the smoking strip*

The New Varroa Handbook

Fig 20 *Waiting five minutes*

Fig 21 *Removing the insert with mites*

Varroa Detection in Brazil with Yugoslavian Acaricide

On Monday August 3rd 1992 Bayer held a press Launch to announce that Bayvarol was now available as the first registered bee medicine in Great Britain. The acaricide is also capable of being used effectively to detect mites. Everyone is advised to follow the manufacturer's instructions.

I have checked all my colonies in the following way: Each hive is opened up in the normal manner using smoke to keep the bees under control. When the brood nest is exposed the brood area is identified (the area of maximum bee activity). The recommendation is to

Fig 22 *Photograph of BBKA Chairman demonstrating Bayvarol on the colonies at the Chelsea Physic Garden for the press at the launch*

apply two Bayvarol strips between the brood frames so as to divide the brood nest into three parts. I use my discretion according to the brood area. After 36 to 48 hours the Bayvarol strips are taken out and the inserts removed for examination.

Clive de Bruyn

Once the Bayvarol strips are removed from their foil protection they should be used immediately. If the Bayvarol is only being used to detect the presence of mites as described above than the same strips can be used a number of times (10 to 15). The strips can be stored for a limited time (30 days) provided they are returned to their original wrapping and sealed. On no account must strips be used more than once for treating a colony. I have had good positive results with one strip for 12 hours on the Isle of Wight, Devon and in Suffolk. Not all the infestations were heavy ones. Despite these findings I advise beekeepers to stick to the manufacturer's recommendations.

Fig 24 *Placing the insert in place*

Fig 23 *BBKA Secretary returning the brood box*

Use of an Acaricide on a Sample of Bees

This method is described in a booklet from DARG **Ref 4**. Two brood frames

Fig 25 *Inserting Bayvarol*

from the colony covered with bees are placed in a nucleus box fitted with a stiff paper insert on the floor, covered with Vaseline. The entrance to the nucleus box is sealed and the bees are trapped inside by a hessian cover. Tobacco smoke is blown under the cover from a smoker. After 15 minutes the frames with the bees can be returned to the hive. The insert can then be examined for varroa mites.

This method can be adapted so that all swarms which you collect can be checked prior to putting them into a hive.

Examination of New Brood Comb

In a study tour of Germany in 1992 newly drawn comb which the queen had laid in was examined against the light. Varroa mites showed up as dark objects in empty cells in the light coloured comb. This technique will not work on old comb. Mites do show up on new comb but I would count it rather as "of interest" than a reliable method.

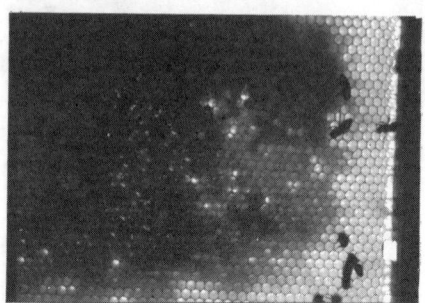

Fig 26 *Holding the comb up to the light*

Discussion

Some of these methods have now been in use now for several years in a number of different countries. Despite this fact there exists no satisfactory

Fig 27 *The view through the comb*

statistical basis for sampling. The main problem is the number of variables that exist. Some of these are:-

1) Differences in hive types and design.
2) Size of the colony.
3) Number of varroa mites in the colony.
4) Time of year.
5) Availability of materials.
6) Labour involved.
7) Analysis treatment available.
8) The skill and experience of the investigator.
8) Variations in temperature and humidity.

A number of different detection methods have been compared in the USA. **Ref 5**. All tests were done in hives known to be infested with varroa. Basically the idea was to obtain some information on the reliability of the different methods. From the results obtained so far It would appear that the use of acaricides are superior to examination of samples of brood or bees using an ether roll or washing fluid. There is still a lack of reliable statistics on efficacy of detection methods which would relate to conditions in the United

Kingdom. Such information will only become available when sufficient data is collected from a vast number of hives or apiaries.

> **Ref 1** Samples marked "Debris for Varroa Search" should be sent to:-
> **The Central Science Laboratory**
> National Beekeeping Unit
> Luddington Nr Stratford on Avon
> Warwickshire CV37 9SJ
> **Ref 2** BBKA NEWS No 70 March 1989 "Supplement No 1"
> **Ref 3** MAFF Pamphlet 936. "Varroasis of Bees, Tobacco smoke detection."
> **Ref 4** Devon Apicultural Research Group. "Living with *Varroa jacobsoni*"
> **Ref 5** American Bee Journal Sept 1089 pp 605.
> "Evaluation of 6 methods of detecting varroa mites in beehives."

3.5 WORLD MOVEMENT OF VARROA

Fig 28 A.C. Oudemans.

It has always intrigued me that this apparently innocuous pest which was, only of interest to a few serious entomologists in the first decade of this century can now be the most notorious parasite that beekeepers have ever known. In less than 100 years this little animal has become known throughout the apiculturel world. I wonder if Mr Oudemans ever realised how famous he would become?

The way varroa spreads naturally would appear to be by movement of bees from one colony to another. Worker bees are known to "drift" between colonies in an apiary. Drones in the active season seem to move freely between apiaries many miles apart. Robbing between colonies is sometimes severe enough for the beekeeper to notice. There is also much robbing that takes place without the beekeeper being aware that it is going on. Either the robbers or the robbed could be infested.

Swarming of infested bees will also enable the mite to enter completely new areas. It would seem likely that natural spread of varroa in an area of low colony density is far slower than in regions of high colony density. How many colonies are within flying distance of your hives?

There is some evidence that mites can drop off foraging bees onto flowers, where they may be picked up by other bees. Adult varroa mites have been sustained on flowers for a more than one day. Varroa mites have been found on several species of insect flower visitors. **Ref 1**. There are also reports of varroa mites being found in wasp colonies. It is doubtful if they could survive for long on wasps. There is no data to suggest that the varroa mites could reproduce outside of a honeybee colony. Any spread due to intermediate hosts must be of comparitivly miniscule importance compared to the mites that are carried on bees directly from one hive to another. Varroa mites are insidious creatures possibly with many methods of dispersing themselves. So far no means have been found to prevent them spreading eventually to all colonies in an area.

Any attempt to accurately chronicle the way that *Varroa jacobsoni* spread from its original host to the mellifera species is full of speculation. The manner in which the mite was spread from one country to another can also only be hypothetical. Despite these limitations it may be useful to look at what has been written down about these aspects of the situation. Such information may be of use in thinking out a realistic strategy for dealing with the problem now that varroasis has been introduced into the United Kingdom.

In my searches through the literature it was soon made clear that the "**first reported discovery**" bore no relationship to the original importation. If one looks at the year in which the mite was first discovered in a particular country only an approximation will be obtained which will be some time after its actual introduction. The time interval between introduction and discovery will vary depending on the state of beekeeping, the level of statutory assistance and research and the concern and diligence of beekeepers. It might be salutary to consider the finding of varroa in England with these criteria in mind?

In the table the year quoted is when the mite was first reported unless otherwise stated. This information is not always consistent and in many instances, especially in the 1940s and 1950s not every source gave the same year. This is not surprising when one considers that at this time the mite was not considered a problem. I am especially puzzled by the fact that it was reported in Japan as early as 1909. It was given a local name by traditional Japanese beekeepers who were managing *A. cerana* in log hives. It is known that *A. mellifera* bees were introduced as early as 1900, yet varroa was not thought to be a serious pest until the 1960s.

Reports began to circulate in 1970 that the parasite was a serious epidemic problem. Apart from some Africanized races of *A. mellifera* in South America we are led to believe that most mellifera colonies will eventually succumb to an infestation. Why did varroa not cause problems in mellifera colonies throughout the first fifty years of this century?

Was it that:
• Beekeepers did not notice?
• Management methods changed, exacerbating the problem?
• The bees prior to 1950 were resistant?
• There was a change in the varroa itself?

The transfer from *A. cerana* to *A. mellifera* in the USSR has been pieced together from verbal accounts. I have not been able to find any origonal papers on the topic despite making some enquiries from Russian beekeepers. The mite was first reported on *A. cerana* in the far East of the Russia. One tends to forget that the old

U.S.S.R. had a foot in Asia as well as Europe. In these Asian forests the local cerana bees thrived in log hives. When Russian peasants from the European side of Russia emigrated to the area they could not get on with the local cerana colonies. *A. mellifera* colonies were brought in from the Ukraine. These first importations must soon have been infested with varroa from the local *A. cerana* bees.

Apparently these mellifera colonies did not all die out. In fact because of the excellent forage and good weather high honey yields were obtained. This led beekeepers in the West of Russia to believe that the bees in the East were of a special "improved" strain. Beekeepers in W. Russia made arrangements to obtain some of these bees. *Apis mellifera* colonies from Asian Russia thus came to European Russia bringing with them their new parasite, *V. jacobsoni*. Before you criticise think carefully about all the beekeepers you know who will insist that some foreign imported bee is better than anything we have here. Undoubtedly in some instances they might be correct. Is it worth the risk? Varroa is not the only disease and parasite that could be imported.

Because of the difficulty of getting information out of the Soviet Union it was not apparent in the West that varroa was becoming a problem.

In the 1970s rumours and reports began to circulate. In 1971 the loss of 55,000 colonies in the Moscow region was attributed to varroa. Once varroa was established in the Western Honeybee of Europe it was only a matter of time before it spread from European Russia to the rest of Europe.

Bulgaria was the next European country to report "Whole districts where many colonies once thrived are said now to be without bees". Reports of heavy colony losses continued to come in as each new area became infested.

The introduction of varroa into Germany is believed by many to be the result of direct transfer from *A. cerana* bees brought into Obersal for research purposes. It is reported that the entry of varroa into South America was due to the importation of queens to Paraguay from Japan. In most instances no one is going to say with any certainty where and how the parasite first arrived. Even countries with strict long standing restrictions on imports were not safe. The discovery of varroa and its subsequent spread throughout the U.S.A. is well documented. It will be interesting to monitor the spread of varroa in the U.K. in the next few years.

Different countries adopted different policies when the mite was first discovered. A ban on movement often resulted in achieving just the opposite. Afraid of loosing their livelihood migratory beekeepers, in some instances, deliberately moved their bees outside the exclusion zone so that they would not be caught in the restricted area. Even without such illegal movement no ban on movement has ever been shown to have any effect. Where restrictions on movement have been applied they have often been politically motivated without any technical basis. A strict authoritarian destruction policy was ruthlessly tried in Czechoslovakia and Mexico. In neither case was it possible to eradicate the mite.

Clive de Bruyn

When the mite is first discovered there is much interest and concern. However, when beekeepers find that their colonies do not die immediately they begin to think that all the panic was unjustified. Later when mite populations do become large enough to cause damage there is a resurgence of interest. After some years there is an adjustment in the bee population and the people who keep bees. Eventually with some changes in management apiarists will accept varroa as just another problem to be dealt with. I see no reason why this scenario should not run its course here.

The experiences suffered overseas can be summarised. Like the proverbial curate's egg some news is bad and some news is good:
1) Even countries with stricter bee importation laws than the U.K. have not been able to exclude varroa.
2) In no country has a policy of eradication of infested colonies stopped the spread of infestation.
3) In no country has a ban on movement done anything to contain the spread of infestation.
4) There is no "cure" for varroasis.
5) Once it arrives in your colonies you and the bees will be at the bottom of the learning curve. Some bees and some beekeepers will not last the course.

I have often wondered about "the spread of acarine from the Isle of Wight." I use quote marks because I do not believe that the mite originated on the I.o.W.. Unfortunately the records from that time do not tell us enough to judge the true story. We now all accept that *Acarapis woodi* is not a serious problem to beekeeping in this country. It is only sometimes a problem for some colonies. I would like to imagine that the impact that varroa has on the way we keep bees up to the start of the new millennium will be carefully chronicled for future generations. Think fellow beekeepers how fortunate you are to live at such interesting times.

Bernard Halstead lives on the Ledge Rd., St. Stephens which is on the boundary between New Brunswick (Canada?) and Main, USA.
October 1990, The first federal apiary inspector arrived to inspect colonies for varroa mites in Charlotte County along the 20 mile border where there were beekeepers. Mites were only found in one apiary during the search (fluvalinate plastic strips for 72 hrs. a sticky plastic tray in the bottom of the hive was then inspected for mites, verified under the microscope.)
The 10 hives in the infested apiary were "depopulated". The beekeeper being compensated at the rate of $60 per hive. He ordered four packages of bees from New Zealand and started beekeeping again.
In November 1991 the inspectors returned and the same 20 mile area was inspected again. This time they found 11 apiaries with varroa mites. These were all "depopulated" including the one restocked with N.Z. bees. Included in these were the owner's bees and 13 hives belonging to the Charlotte County Beekeeping Association and a neighbour with 34 all purchased that year suffered the same fate.

Bernard Halstead
RR3 Ledge Rd. St Stephens N.B. E3L2Y1 New Brunswick

FIRST SIGHTINGS OF VARROA

Java	1904	First reported Oudemans *A. cerana*
Japan	1909	Reported *A. cerana*
	1953	First economic impact noted (2)
Sumatra	1918	*A cerana* (3)
China	1951	Reported *A. cerana*
	1957	Reported *A. mellifera* (4)
Singapore	1944	Reported (5)
Korea	1950	Reported (6)
Pakistan	1955	Reported (5)
Hong Kong	1962	Reported (5)
Philippines	1963	Reported (7)
USSR (E. coast)	1964	Reported (5)
Bulgaria	1970	
Yugoslavia	1972	
Paraguay	1973	On queens from Japan?
Argentina	1976	
Turkey	1977	(5)
Rumania		
German Federal Rep.	1977	From Pakistan five years previously
Brazil	1978	S. Paulo from Paraguay
Uruguay		
Greece		
Tunisia	1980	From Rumania (5)
Finland		USSR Border
Algeria	1981 (5)	
Italy		
Poland	1982 (8)	
Holland	1983 (9)	
Israel	1984 (10)	
Iran		From Turkey (11)
Spain		
Portugal	1987	Serpa on the border with Spain
USA		September 25th/28th 1987 (12) Migratory beekeeper Wisconsin
Morocco	1989	Aug
Canada	1990	From USA.
England	1992	April 4th
Mexico	1992	May 3rd Imported queens?

Ref 1 Kevan, P.G. Laverty, T.M. & Denmark, H.A., Association of *Varroa jacobsoni* with organisms other than honey bees and implications for its dispersal. Bee World 71: 119-121. 1990
Ref 2 Tetsuo Sakai & Kazuo Takeuchi (1980). *Varroa jacobsoni* and its control in Japan. XVI Int. Cong. of Ent. Kyoto, Japan 3-9 August 1980
Ref 3 J. Walker. Report of a seminar on varroasis Bucharest, Rumania Aug 20-26th. personal communication.
Ref 4 Zheng-You Fan & Lung-Shu Li (1988) In: Africanized Honeybees and Bee Mites, Ellis Horewood Ltd Chichester, W. Sussex. England: pp 415-419,
Ref 5 Bowman C.E., Griffiths D.A., (19). World distribution of the mite *Varroa jacobsoni*, a parasite of honeybees
Ref 6 Seung-Yoon Choi (1988) In: Africanized Honeybees and Bee Mites, Ellis Horewood Ltd Chichester, W. Sussex. England: pp 413-416,
Ref 7 Delfinado M. D. (1963). Mites of the honeybee in S. Asia. J. Apic.Res. 2:113-114
Ref 8 Kosonocka L. (1989). Varroa in Poland. A.B.J. Sept 1989. pp 597-599
Ref 9 Tiggelaar H.M. personal communication.
Ref 10 Gerson U. et al (1988) In: Africanized Honeybees and Bee Mites, Ellis Horewood Ltd. Chichester, W. Sussex. England: pp 420-424,
Ref 11 Komeili A.B. (1988). The impact of the varroa mite on Iranian commercial beekeeping. A.B.J. June 1988. pp 423-424
Ref 12 Lawrence C. (1989) The proposed varroa mite program. A.B.J. May 1989. pp 309-311.

3.6 SPREAD IN THE U.K.

THE FIRST DISCOVERY IN BRITAIN
by Ron Brown

On Saturday afternoon 4th April 1992 the Cockington, Branch Apiary, Torquay went down in history as the official location where the mite *Varroa jacobsoni* was first found in the United Kingdom. It came about as follows:-

Like all good beekeeping citizens the Torbay branch forwarded floor debris for mite examination to the National Beekeeping Unit. A letter had already been received from the N.B.U. saying that no mites had been found. The branch had arranged a meeting to show members the M.A.F.F. technique of using tobacco to test for varroa mites.

Our Apiary manager, Jack Berry, consulted me as to whether there would be any point in carrying on with the demonstration in view of the "All Clear" message from the Ministry. We discussed the matter and I expressed my strong feelings that we should continue. At least we should get a check on any *Braula coeca* present"

In the event not many turned up on the Saturday. (Typical, Editor). A group of four hives were smoked as described in MAFF Pamphlet 936 **(1)**. After about an hour the entrances were unblocked and the brood boxes were lifted off for a moment so that the paper inserts could be removed. A cursory examination showed small pieces of wax, excreta and other debris plus several easily identified braula as expected.

The paper inserts were carefully folded and taken back to Jack Berry's home for examination. About an hour later I received a 'phone call from Jack saying that Margaret and Rod Saffery had found what they believed was a varroa mite and would I come over to check. There was no doubt it was a varroa mite.

The paper inserts were examined most thoroughly, using hand lenses, but no further mites were found. The four hives were again subjected to tobacco smoke and this time the inserts were left in overnight. At 9.30 am on the Sunday the inserts were removed. At the same time the only patch of sealed drone brood (purple-eye stage) was uncapped. A quick visual check with the naked eye revealed some possible mites.

Back at Jack Berry's house the inserts were spread out on a large table. Over a period of two hours every square inch was studied with magnifying glasses. Several individual mites were transferred to microscope slides for confirmation at high magnification. In all nearly 20 mites were positively identified. Some had been knocked down by the tobacco smoke and others had been found on the decapped drone pupa.

(1) Varroasis of bees, Tobacco smoke detection. MAFF Pamphlet 936, 1985

Following the "first official" discovery on the Saturday the Ministry were informed. On the Sunday the Ministry National Beekeeping Adviser, Medwin Bew arrived to co-ordinate the work of the Bees Officers. With the help of local B.D.O.s, arrangements were made to check all stocks of bees within a 3 mile radius of Torbay.

On Monday 6th the BBKA general sec wrote to all Association secretaries advising them of the finding. Under article 8(1)(c) of the Bee Disease Control Order of the Bees Act an infested area was declared by MAFF on the 7th of April restricting movement of honeybees for a 6km radius around the infested apiary at Cockington. This first discovery appeared to be only a light one suggesting that it had spread to this apiary from a more heavily infested source. It was thought that the infestation was one on the periphery of the outbreak rather than the centre.

Further investigation of bee stocks in Devon disclosed more infested apiaries (7) and restrictions were extended. Most of early infestations found indicate that the mite has been present in colonies for at least one season in some cases the infestation levels indicated that the mites must have been here for several years.

Form 9

Article 8(1)(c)
(Infected Area)
BEES ACT 1980
BEE DISEASE CONTROL ORDER 1982

NOTICE OF DECLARATION OF INFECTED AREA (No.2)
The Minister of Agriculture Fisheries and Food in exercise of the power on him by article 8 (1) (c) of the Bee Disease Control Order 1982 hereby declares the area described in the schedule to this notice to be an infected area for the purpose of that order and the following provisions shall apply-
1) No hive, bee, comb, quilt, bee-product or appliance may be moved into, from, or within the infected area except under the authority of a licence issued by the minister;
2) Every owner and every person in charge of a hive, bees, combs, quilts, and bee products or appliances in the infected area shall (as soon as he is reasonably able to do so) notify the Minister of Agriculture, Fisheries and Food in writing of his address and the location of the hives, bees,

The New Varroa Handbook

> combs, quilts, bee-products or appliances belonging to him or in his charge.
> FAILURE TO COMPLY WITH THIS NOTICE MAY CONSTITUTE AN OFFENCE AGAINST THE
> BEES ACT 1989
>
> SCHEDULE
> The infected area is all that land found within a radius of 40 kilometres of St. James Church, Oakhampton in the county of Devon.

On 14th April MAFF arranged a meeting of representatives of British BKA, BFA, Welsh BKA, Scottish BA, CONBA and NBU.

It was accepted that varroa was indeed present in Britain and it would not be possible to eradicate it. MAFF would not be ordering a general destruction of infested colonies. MAFF would be making arrangements for a supply of Apistan strips to be granted a special licence. These would be offered to beekeepers with infested hives. The special licence would be in the name of the National Beekeeping Unit who would direct their use through local BDOs. Anyone using these chemicals to treat their bees would not be allowed to pack and use any honey from the treated hives.

> ### THE FIRST DISCOVERY IN THE ISLE OF WIGHT
> **by Mary Case**
>
> My earliest opportunity to scrape hive floors in 1992 was on Easter day April 19th. I sent the scrapings from each of my apiaries to Luddington and waited for the all clear. At about 9.30 on 23rd of April I had a telephone call from the Bee Disease Officer who said he had some bad news for me. The N.B.U. had found varroa mites in 4 of the apiary samples. After the initial shock I suggested that the infested colonies should be destroyed. However, he told me that M.A.F.F. did not recommend destruction.
>
> The local I.W.B.K.A. secretary was told that varroa had been found in the S.W. of the Island. Within 10 minutes my phone was ringing again, the chairman of the I.W.B.K.A. rang to inform me that varroa was in my area. I had a split second to decide to tell him that it was my bees that had varroa. I wonder what would have happened if I hadn't owned up?
>
> Three weeks after varroa was confirmed on the island the local BDO arrived. We smoke tested all the hives in the area. MAFF was not able to offer much help in treating the infested stocks other than offering Apistan acaricide strips. Luckily for me John Cossburn was still the CBI for Hampshire at the time. He was a great help and soon I was regularly removing sealed drone brood to reduce the build up of mites.

Clive de Bruyn

> How varroa came to be found on the S.W. side of the island remains a mystery. I have no links with Devon bees and it is considered too costly to take colonies to the mainland for the heather.

Beekeepers were given an update from Medwin Bew at the BBKA Spring convention at Stoneleigh on April 25th. It appeared at that time that the original epicentre of the infestation would be on the heather moors of Dartmoor. Severely infested colonies with 1,000s of mites had been found indicating that the mite must have been present for several years. It was known that these moors had been visited by migratory beekeepers from as far away as Kent and Yorkshire within the last five years.

Government policy, we were told, would be in line with previously announced contingency plans. There would be no compulsory destruction of infested colonies nor would there be a ban on the movement of bees. It was thought that it would be virtually impossible to enforce any legislation to prevent beekeepers moving hives. Natural spread of the mite and movement of hives by beekeepers who were not aware of the mite must have spread the mite already. In addition the Ministry did not wish to damage the livelihood of those beekeepers who are dependant on the movement of hives for a crop.

MINISTRY OF AGRICULTURE FISHERIES AND FOOD

CONTINGENCY ARRANGEMENTS FOR DEALING WITH AN OUTBREAK OF VARROASIS IN HONEY BEES IN ENGLAND (REVISED 1989)

2. PROPOSED ACTION
(viii) If an outbreak is found to be isolated and eradication is considered practicable, all colonies in the apiary would be destroyed. In all other circumstances on present technical knowledge there would be no benefit from applying standstill notices because it would have to be accepted that the disease was widespread throughout England. England would therefore be declared an infected area and ADAS would concentrate its efforts on advice to beekeepers and research, on detection and control based on knowledge of the effectiveness of measures taken on the Continent.

It came as somewhat of a surprise therefore to find that on the 27th April the Ministry had declared a new infected area. At that time no varroa mites had been found in all these counties. Despite an intensive search by local beekeepers and the ministry BDOs no varroa were found in Kent.

The New Varroa Handbook

> ### INFESTED AREA 27th APRIL
>
> Following confirmation of the presence of varroasis in a number of colonies in the south of England a statutory infested area was declared south of a line from the Severn to the Thames. There is no restriction of movement within the infested area, but it is illegal to move bees and hive frames containing comb into or out of the infested area without a licence. The counties included in this area are:-
>
> CORNWALL, DEVON, AVON, DORSET, SOMERSET, WILTSHIRE, HAMPSHIRE, BERKSHIRE, BUCKINGHAMSHIRE, EAST & WEST SUSSEX, ISLE OF WIGHT, SURREY, OXFORDSHIRE, KENT, AND GREATER LONDON.

This decision could not have come at a worse time. Migratory beekeepers had already moved colonies into Kent for pollination. In addition the MAFF Regional Administrative System was in the midst of a complete reorganisation. From April 1st the old local county ADAS offices were consolidated into 9 Regional Service Centres. This meant that beekeepers in Kent had to now refer to a regional office at Reading.

On 14th May David Curry, Minister of State at the M.A.F.F. met with representatives of the beekeeping organisations. He decided that special arrangements should be made for hives temporarily in Kent for pollination.

"To deal with a particular problem now that the pollination season is ending I have agreed a general movement licence out of the infected area for hives temporarily in Kent, subject to the need to inform my Department of the destination of the hives. This licence will last from one month from today (May 15th). I feel able to do this since there is no evidence of varroasis in Kent. For the rest of the infected area beekeepers will need to apply to their Regional Service Centre for a specific licence to move."

Varroa continued to be found through the season. (See Table at End.) The first instance in Wales was at South Dyfed in a place called Marros. Two beekeepers originally of the DBKA Plymouth branch had been trying to set up a bee farm in Wales. They started their apiary with 12 hives three of which had been on Dartmoor two years earlier. When the hives were tested for varroa, after removing the supers in August. Varroa mites were found in two of the colonies.

> ### FIRST DISCOVERY IN WALES
> ### by D. Williams & P. Phillips
>
> We had 12 hives on our land in S. Wales. Three of these had been on Dartmoor two years previously. At the end of August, after the supers were removed, we tested for mites. In two colonies we found varroa. MAFF were

Clive de Bruyn

immediately informed. A standstill order was placed on the apiary and possible contact colonies in the area were examined for mites. None were found. We also examined our hives in Devon which were clear.

Due to the hysteria aroused by the local press** we decided to destroy the two infested hives and three others in the same apiary. We felt that if we did not take action someone else would have done it for us. The feelings of prejudice and intolerance ran very high. The infestation appears, so far, to be an isolated case. We applied for and were granted licences to take the remaining hives to Devon, even though some were obtained in Wales. The Ministry have tested all these hives and no mites were found. Other beekeepers in Wales have offered to replace hives we destroyed. The Pembrokeshire Branch have been particularly helpful.

We tried to put the journalists right by telling them the facts, but little of this has been printed. We have not even met some of the beekeepers who have attacked us in the press. This was all very distressing as our first concern was for other beekeepers in the area:- to be branded as "The Nasties". We had done all we could to comply with the law and warn other beekeepers in the area.

Some of the newspaper articles were very derogatory and bordered on racism. The implication was that we had brought varroa into Wales illegally. Who will have the courage to look for, let alone report a case of varroa if they are treated as we were?

**
- English Couple Brought Bee Bug to Wales
- "Beekeepers say infected hives should be burned"
- Diseased Hives to be Destroyed.
- "Smuggled into Wales by an Englishman"

There was a great deal of very sensational newspaper publicity which caused some alarm locally. Contrary to MAFF advice the Phillips destroyed the infested hives and returned the rest to Devon after obtaining a licence from the ministry to do so. The Welsh beekeepers generously offered to replace all stocks

During October MAFF organised an intensive search in areas outside the declared infested area where it was thought probable that there might be an infestation. Infestations have been found north of the statutory "infected area" of southern England. Mites were found in a single apiary in Lincolnshire during the first few days October. These could be related to colonies moved unwittingly from Kent after pollination. Colony collapse directly associated with varroa have already occurred in N. Dartmoor. A light infestation was discovered in Kent in late September. Further checks in the area brought to light mites in seven apiaries. 13 infested apiaries have also been discovered in Suffolk.

113

THE FIRST DISCOVERY IN KENT
by Brian Stenhouse

On 2nd September a swarm entered an empty hive at 5.00 in the evening. The bees appeared healthy and the queen was soon laying well. On 17th September a screened varroa floor was fitted and a single Bayvarol strip inserted. The check for varroasis was positive! Following this find an intensive check round was made on other sites. In all instances 1 Bayvarol strip was inserted in the centre of the brood nest for 24 hrs. Half the hives had a screened varroa floor and the remainder had an insert laid directly on the floor. Not all the colonies on each site showed mites. This could indicate that only testing a few colonies in each apiary is not reliable. (Using two strips for a longer period may have knocked down more mites, Editor)

The heaviest knock down occurred in the swarm (45 mites). In most of the established colonies only one to three mites were knocked down (max 7). At the end of September a number of apiaries in the Faversham area all showed infested colonies. Varroa mites were found in a colony brought in from Northumberland for breeding. This stock had been isolated from other colonies and had not been moved since it was brought down in April 1992. One must presume that it had picked up varroa since its arrival in Kent. (If this is the case then it gives encouragement that mites can be detected comparatively early if the proper techniques are used, Editor).

One colony has had a screened varroa detection floor in place since July 1992. This had been monitored regularly. Just prior to inserting the Bayvarol strips after finding mites in the swarm one single mite was found on the varroa floor. Presumably this was natural mite mortality. The colony was treated with four Bayvarol strips. At the end of the first week only nine mites were knocked down.

There have been no movements of bees out of Kent at all. In 1992 colonies were moved within the county. There have been no importations of queens from anywhere in the UK for at least the last five years.

THE FIRST DISCOVERY IN SUFFOLK
by John Blakesley
Bees Officer

MAFF Bees Officers carried out searches in South Suffolk and North Essex with tobacco smoke in Autumn 1991. No varroa were found. It was therefore somewhat of a surprise to find an established infestation in Felixstowe on 13th October 1992. This incident does not appear to be connected to any other infestation in the country. It could have come directly from the continent on the many freight movements into Felixstowe docks, possibly 4 to 5 years ago. The hives with the highest mite counts were all close to the docks. The beekeepers in the area are all hobbyists, most with static sites.

No apiary checked in the Felixstowe area was found to be free of varroasis. It was discovered during the inspections that beekeepers from Ipswich had inadvertidly moved bees close to Felixstowe for oil seed rape and pollination. Subsequent checks in Ipswich showed the presence of varroa. Some static apiaries were also checked and low levels of mites were found. Two sites across the river Orwell from Felixstowe on the Shotley peninsular were also found with low levels of varroa. An apiary near Clare was found with a low level of mites in early November. Ministry checks were then concluded until 1993.

The mite counts were
    Felixstowe   36 colonies   14494 mites
    Ipswich   48 colonies   548 mites

The apiary which had the highest mite counts was monitored over a period of time. The apiary was tidy with clean hives, well maintained. At the first inspection about 40 dead mites were found on each floor. Outside the entrance on the ground dead mites and deformed bees were noticed. Bayvarol strips were inserted in the brood nest, to detect mites, according to the manufacturers instructions. An insert made up of a 15" sticky plastic (Fablon) square covered with greenhouse shading (3mm Netlon) was placed on the floor to catch the mites. After 24hrs the inserts were removed and the screens inspected. The average mite count per hive was about 1,000.

The Bayvarol was left was left in one of the colonies to check how effective it was in treating the infestation. After the six weeks the first Bayvarol strips were removed and a fresh set inserted for a further two weeks.

	wk 1	wk 3	wk 6	wk 7	wk 8
Mite No.	600	420	10		7

The Bayvarol was then removed. The bees will be monitored again in 1993. At the start of the treatment none of the colonies had any drone brood. Worker cells which were opened had 4-5 mites in each. At the end of the treatment the cluster had reduced from 6 combs on October 13th (1 frame had a 3" circle of brood) to 4 frames on December 1st. After six weeks of treatment 15 sealed cells were opened but no varroa mites were found.

The New Varroa Handbook

VARROA IN IPSWICH
by David Little

In late October my apiary of 8 hives was checked for varroa mites using Bayvarol strips for 48 hours. Two hives showed varroasis. One mite and eight mites were found on the Fablon inserts. My feeling is that we have been waiting for this for 10-15 years:- now we must live with it.

The two infested colonies were treated with Bayvarol for six weeks. All colonies will be checked in 1993. I intend to monitor the mite levels by checking drone brood in the summer. I will treat with Bayvarol if I consider the mite levels are excessive.

It can be seen that varroa can be found by anyone who looks. Some of the first discoveries have been made by BDOs others have involved ordinary beekeepers or noted personalities. All that is required is that one looks. The techniques used have also been different illustrating that reliance should not be placed in using only one method. Different techniques are more reliable at special periods in the year.

The result of these discoveries in Autumn 1992 resulted in another Ministry line. Aside from the imposition of another line the way in which the information was put out was not very satisfactory.

In answer to a parliamentary question from Tam Dalyell MP (Dinlithgow) Mr Curry said:

"To contain the spread of varroa, and in consultation with the beekeeping organisations, we declared an infected area south of a line between the Thames and the Severn. Movement outside the area is prohibited except under licence. Up to now the disease has been confirmed in 267 apiaries within the infected area. Varroa has recently been found in 17 apiaries in Suffolk and one in Lincolnshire. Following further consultation with the beekeepers, I have decided that the infected area will be extended from midnight tonight to include the counties of Bedfordshire Cambridgeshire, Essex, Hertfordshire, Lincolnshire, Norfolk and Suffolk. I have written to the beekeeping organisations and instructed that copies of the letters should be placed in the library of the house"

MAFF extended the area to include E. Anglian counties of ESSEX, SUFFOLK, NORFOLK, and CAMBRIDGE BEDFORDSHIRE, HERTFORDSHIRE as well as LINCOLNSHIRE.

Personally I think that the above information should have been printed in full in all national press. Paid for by the government if that is the only way they could get press coverage. Such a document in the library of the house is

Clive de Bruyn

not likely to be noticed by the non association beekeeper who subscribes to no beekeeping publications.

Policy for future based on the following assumptions

Eradication of *Varroa jacobsoni* from the UK is not feasible. Restriction of movement may retard the spread of varroasis. Nothing will prevent it eventually spreading to every colony. The present distribution of varroa in the UK is not known.

What legacy are we leaving for the next generation of beekeepers?

Discoveries of Varroa infested apiaries in 1992

COUNTY	Apr	May	Jun	Jul	Sep	Oct	Nov
Avon			1	1	1	1	1
Bucks			1	1	1	1	2
Cornwall			1	1	1	1	1
Devon	36	42	54	56	64	74	77
Dorset		1	3	3	1*	2	4
Dyfed Wales			2		1	1	1
E. Sussex							2
G. London				2	2	2	3
Hampshire	1	2	6	7	8	10	17
Isle of White		6	6	6	6	7	7
Kent					2	7	12
Lincolnshire						1	1
Somerset	13	18	18	19	37	40	41
Suffolk							17
Surrey	6	9	18	24	31	46	76
W. Sussex			1	2	2		2
E. Sussex						2	2
Total	56	78	111	122	157	195	287

MAFF Statistics

The New Varroa Handbook

VARROA IN UK

A - B
Statutory infected area 27th April 1992

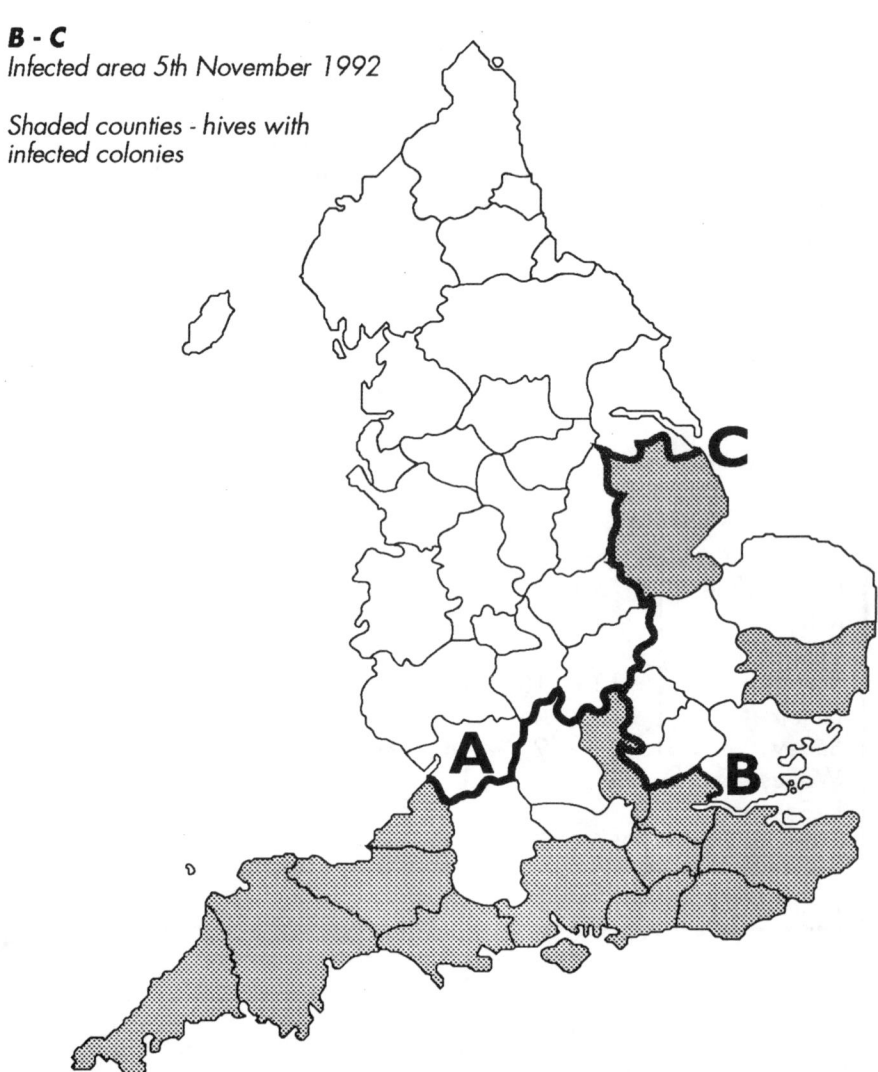

B - C
Infected area 5th November 1992

Shaded counties - hives with infected colonies

Clive de Bruyn

MEDICAMENTS 3.7

The use of chemical acaricides to combat varroasis will probably be necessary in the immediate short term. To maintain a reasonable number of productive colonies most beekeepers will have to resort to such measures. However, such treatment should not be undertaken without some appreciation of the consequences.

Against:

1) Risk of contaminating bee products with residues. This can occur from a single application or over a period of time by continuous use of the same chemical.
2) Risk of the parasite developing resistance to the chemical. Continuous use of the same medicament will increase the possibility of resistance developing.
3) Experience is needed to get the timing and application right to do most harm to the parasite without damaging the colony and hive products
4) Cost.
5) A reliance on chemicals will delay the development of a stable host - parasite relationship.

Chemical Residues

There is a great deal of concern about the use of chemicals to treat infested colonies and the concomitant problems resulting from residues in honey. Even careful use of chemicals can, with repeated use over a number of years, cause residues to accumulate inside the hive. Irresponsible use will increase the risk of leaving residues.

Residues from unauthorised treatments can create a great problem. Such residues would have a serious impact on consumer confidence in the product. There is considerable public concern at present regarding residues in food products. Analytical procedures are now capable of virtually finding a little bit of anything in everything.

Under the Food Act two important criteria are considered:
a) The maximum allowable residue
b) The maximum daily intake

We are all coming into contact with chemicals every day. For example: In the bathroom; shaving soap, aftershave, toothpaste. At breakfast; coffee, toast, orange juice, In the house; washing up detergents, floor polish, air freshner, Through one's hobbies; garden centre chemicals, alcohol, tobacco.

What each of us must decide is what is an acceptable risk? There must be a balance between the benefits and the risks. It is not easy to resolve even with lots of data. Consider the odds. Make your bets.

The New Varroa Handbook

Public concern about the purity of the food it consumes has never been greater. This has resulted in more stringent government surveillance coupled with stricter controls and legislation. Any product or substance which is manufactured, sold or supplied for the purpose of treating ectoparasites or endoparasites which live on bees is regarded as a veterinary medicine and therefore requires licensing in accordance with the provisions of the Medicines Act.

The Medicines Act 1968. specifies that no medicinal product may be sold supplied or imported into the UK without a product licence. The Veterinary Medicines Directorate (VDM) is the body to licence chemicals for use in the UK. The registration procedures are there to safeguard the public against the use of unnecessary or undesirable products. Companies wishing to market such products in the U.K. must seek to obtain a product licence under the Act.

Applications can be made to the Medicines Unit of the Central Veterinary Laboratory, New Haw, Waybridge who will be able to advise companies on any technical matters relating to their product licensing application.

Since varroa first appeared in Europe a great deal of research has been carried out for substances which would be effective against the mites yet do no damage to honeybees. In almost all countries of Western Europe bee medicaments must meet the requirements of the local VDM. Possible toxic and mutagenic effects upon humans are of particular importance for the process of registration.

Besides the toxicological studies, tests must be carried out for residues in bee produce and possible danger to consumers. The widespread use of a substance in one EEC country will not guarantee its registration in another. Substances registered in one country are not necessarily examined under the same set of criteria in another. Currently each have their own set of values and priorities.

The decision to register a product against varroa in Britain is purely a commercial one made by the companies themselves. From the manufacturers point of view the market in Britain is considered relatively small. The Ministry of Agriculture together with BBKA and BFA have approached a number of manufacturers to encourage them to register their products in Britain as it would be desirable to have a range of effective chemicals to use against varroa.

Ever since the 1950s pest control specialists have been aware that a single chemical approach to crop protection is not a reliable answer. At this time of writing there is only one approved chemical for the detection and control of varroa mites in Great Britain.

Alternating the medicaments used in control will reduce the risk of the mites acquiring resistance to particular chemicals. Through the use of a battery of approved chemicals the onset of resistance building up can be delayed significantly. Remember that chemicals which are similar in their mode of action may eventually all be tolerated by the parasite, as the mode of resistance may cover the entire range of similar products. For Example Apistan/Bayvarol (both artificial pyrethroids).

The aim of any management must be to keep varroa populations low enough so that the honeybee colony can live with the infestation. The possibility of eradicating the disease is not realistic. The ideal medicament should kill all the mites, not harm the bees in any way and leave no residues in any products of the hive. No such material has appeared so far.

As bees actively clean the hive of all kinds of "pollution" and regulate the inside of the hive to a constant temperature, CO_2 and humidity. Any foreign substance such as the introduction of an acaricide is almost immediately counteracted by the bees. Therefore the effect of some chemicals is limited to a very short period after application. To increase the efficiency treatments are usually repeated several times. This increases the chances of residues appearing in the products.

Applying a lethal dose of an acaricide to a varroa mite without harming its host is not easy. In the active season a high proportion of the mites are protected in the sealed brood cells. The number of mites in a colony also alters through the year. Some chemicals work by evaporation. The temperature, size of the hive and the number of bees will effect the rate at which such products work. Other chemicals need to be dispersed by the bees so the size of the colony is relevant. These are just a few of the complications.

The medicines available can be divided into hard or soft chemicals. The distinction between these is not always clear. Hard chemicals are generally defined as toxic man made compounds with a relatively long life. Soft chemicals should therefore be "less toxic" "natural products" which break down into innocuous elements fairly quickly. Any sixth form student of chemistry would soon point out that such a classification would have no validity to a scientist. I hope that you and I can understand each other.

HARD CHEMICALS

Anti Varroa Amitraz

This product is registered in France. Application requires a special thermal aerosol applicator. First the emulsifiable concentrate is diluted in water (20ml in 1 litre). The air in the applicator is preheated to 35-40°C. The aerosol is injected into the winter cluster. (Outside temperatures >6/8°C are recommended) for 1 to $1^1/_2$ minutes. The temperature of the air/aerosol mixture is sufficient to break up the cluster getting the chemical to all the mites in the hive. A second treatment three/seven days later is recommended. Application is labour intensive and time consuming. There is a fear of suspected carcinogenic effects. A gas mask is recommended when treating colonies.

In Israel the control of Varroa between 1984 and 1987 was based on frequent fumigation of infested colonies with Amitraz at 3-4 day intervals **Ref 1**. Despite this treatment about 40% of the colonies perished in 1986. Problems were also reported that the Amitraz caused colonies to supersede their queens. **Ref 2**

Apistan

This product is the recommended treatment for varroasis in the USA. It is one of the simplest and most convenient ways of introducing an acaricide into a hive of bees. An impregnated PVC resin carries the active ingredient, Fluvalinate**. The Fluvalinate acts primarily as a contact poison by disrupting the central and peripheral nervous systems of mites and other arthropods.

Two strips per brood chamber are recommended. (One strip between frames 3-4 and one between 7-8). The principal dispersal of Fluvalinate in the colony is by the bees contacting the Apistan strip and then from bee to bee until eventually the varroa mites come into contact with the chemical on the bee's body. The manufacturers claim that in colonies with little or no brood more than 90% of the mites are killed in 5-7 days. In colonies with capped brood 99-100% control is generally achieved in 5-6 weeks depending on season and infestation level.

It is strongly recommended not to use the Apistan strips more than once. Whilst the active ingredient may still be effective in killing some mites the level of Fluvalinate will have been reduced. Varroa mites exposed to low medicament levels may not be killed. Mites which are tolerant to this reduced level of the chemical will be selected for. Any resistance to the chemical will soon spread through the mite population.

Disturbing information is coming from America on Fluvalinate being absorbed into beeswax **Ref 3**. Being fat soluble, Fluvalinate can be absorbed into wax and honeybee cuticle. Gas chromatography analysis confirmed that the longer the strips are left in the hive the more residue was present in the comb and the bees. Residues were not found in the honey. Fluvalinate as an artificial photo stable pyrethroid pesticide does not readily break down. For this reason it would seem prudent to not exceed the period laid down by the manufacturers with regard to treatment.

Recent work advocates that the impregnated strips can be effective if they are applied other than as recommended by the manufacturer. It has been argued that the absorption of the chemical by the wax would be

---

** Originally a synthetic pyrethroid insecticide developed for the control of pests in cotton crops.

reduced if the strips were used on top of the frames or at the entrance. If the material could be made with a series of holes through which the bees would have to pass it might also help to reduce infestation from neighbouring colonies. **Ref 4**. Such experiments against the manufacturers precise recommendations would be illegal under our laws.

The active ingredient, Fluvalinate, is present in other agricultural formulations used for horticultural crop treatment. In France Fluvalinate is obtainable in Klartan. Maverik Aquaflow™, made by Zoecon. has been used in Israel **Ref 2**. I know that beekeepers in these countries have been making up solutions of these chemicals and impregnating wooden and cardboard strips.

Wooden strips soaked in the chemical are not the optimum treatment for varroa mites. Changeable texture in the wood make the efficiency and safety of the wood strips impossible to maintain. The wood absorbs the chemical at different rates which results in a variable rate of release. The strips can also vary in the period over which the chemical is released. Such uncontrolled haphazard use of these illegal compounds can only be deplored.

Apitol

A water soluble systemic applied in sugar syrup trickled between the combs. The bees lap up the sugar and spread the chemical amongst the workers of the colony. Two applications at an interval of seven days are required. The outside temperature should not fall below 10°C. as there can be mortality amongst the bees due to the chilling effect of the liquid.

Bayvarol

The only chemical at this time of writing licenced for use against varroa in colonies of honeybees in Great Britain. (released on Aug 3rd 1992). This product together with Apistan are the latest generation of anti-varroa

Fig 29 BBKA Exec. Chairman, Eric Fenner Inserting a Bayvarol strip into a colony at the Chelsea Physic garden at the Bayvarol launch on August 3rd 1992

chemicals. Both products make use of an artificial pyrethroid impregnated into a plastic.

Flumethrin is the active ingredient in Bayvarol. The mites are killed by coming into contact with the medicament which is picked up by the bees as they walk on the plastic. In a short time the acaricide is distributed throughout the colony. The treatment is fairly long term so that mites can be killed as they emerge from the brood cells.

The New Varroa Handbook

The standard method of application is by inserting Bayvarol strips into the spaces between the combs in the central brood-rearing area in such away that they can be occupied by bees on both sides. The tabs are bent outwards at the marked fold lines and hooked over the top bar of the frames (see **Fig 30**)

Medium strength colonies receive four strips (standard dose). Nuclei and small colonies need two strips ($^1/_2$ standard dose). Strong colonies occupying a double brood chamber need four strips per chamber (2x standard dose). For large colonies occupying several brood chambers two strips can be joined together end-to-end. (see **Fig 31**)

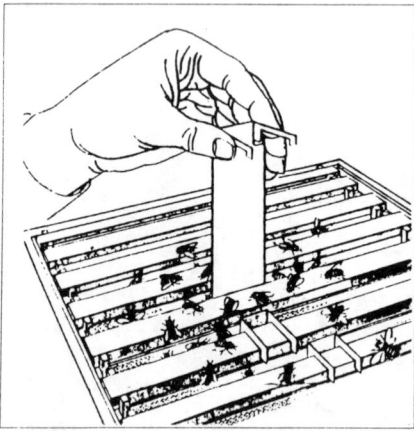

Fig 30

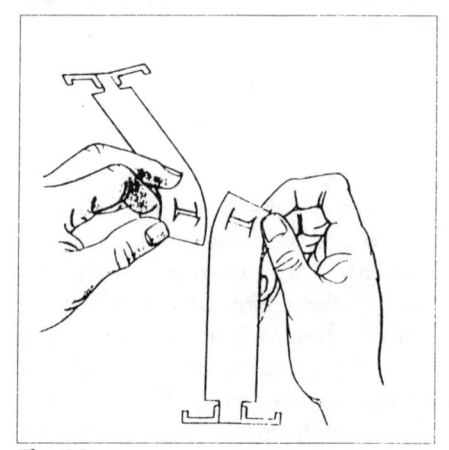

Fig 31

Fig 32 *Two strips in place. Two more to go*

The strips are most efficacious if they are used at the end of the season when all the honey has been extracted and the brood area is being reduced. Use of Bayvarol on honey production stocks is not recommended during a flow. The manufacturers state that the strips must be left in the colony for six weeks. When the treatment is over the strips plus the residual acaricide in the plastic must be removed. This reduces the likelihood of sub lethal acaricide concentrations remaining in the colony which could promote the development of resistant mites. Safe disposal of used strips is by wrapping in paper and placing with domestic waste for refuse collection.

Bayer recommend that operatives using Bayvarol should not eat, drink, or smoke whilst handling the material. It would also seem sensible to wash one's hands after handling the impregnated strips.

Clive de Bruyn

It is probably not appreciated by beekeepers that strictly speaking, as an animal medicine the use of Bayvarol requires records to be kept. The legislation is obviously intended primarily for farmers with traditional livestock.

The records required are the name of the medicine, supplier, date of purchase, batch number, quantity and use (date of treatment, duration, colony i.d.,) and disposal. Nothing too difficult in my opinion. These records would also be important should there be any come back regarding residues in any of the hive products offered for sale. I see no reason why such records should not be kept.

Folbex VA

One of the earliest chemicals available in Europe. The active ingredient, bromopropylate is applied on a fumigant strip. It is recommended that the Folbex is applied to the colony four times at four day intervals. The minimum temperature for its effectiveness is 8°C. Application should be made to the hive in the evening when there is no brood. (Folbex VA cannot penetrate sealed brood). Unfortunately Folbex VA is notorious for causing residue problems.

It has been revealed that of 112 honey samples collected in the Tubingen region of the German Federal Republic, 33 contained bromopropylate at up to 139µg/kg and up to 59µg/kg for the BBP. The large amount of residues present in the wax are disturbing. The figure of bromopropylate residue in beeswax is 182,000µg/kg. in Belgium. In the past few years Folbex has gone down in popularity mainly due to the residue problems and the high costs.

Malathion

Usually applied as a powder, 0.04gr (5%) powder per hive which is dusted onto the bees on the comb 3-4 times at 7 day intervals. The ambient temperature should be between 15°C and 30°C. Another technique is to apply the malathion in pollen supplement late in the season **Ref 5**. Dry pollen supplement is mixed with malathion to give a concentration of 50ppm. This is placed into empty cells, 200g per frame. These combs were placed in the colony after the cluster had formed. It was noted that in attempting to pack the material in the cells worker bees became contaminated with the product. This would obviously allow mites to come into contact with the chemical. The treatment was most effective when the comb of pollen was placed in the centre of the cluster. Tests showed that varroa mites could be destroyed to a satisfactory level in winter after the cluster was formed.

In 1990 on a walking holiday in Greece I discovered some hives. After a few enquiries I eventually tracked down the beekeeper. Despite the language barrier he was able to tell me that all his colonies were infested with varroa. He did not seem to be worried about varroasis although he had lost ³/₄ of

125

his colonies when the mite was first detected. His really serious problem was chalk brood. He told me he was able to control the varroa with malathion. This was applied regularly as a dust onto the combs covered with bees. He also applied it as a liquid spray. Like all beekeepers when meeting a fellow practitioner of the craft he presented me with a jar of honey. I haven't eaten any of it!

Perizin

This product is widely registered in Europe. It was the first systemic acaricide registered for varroa control

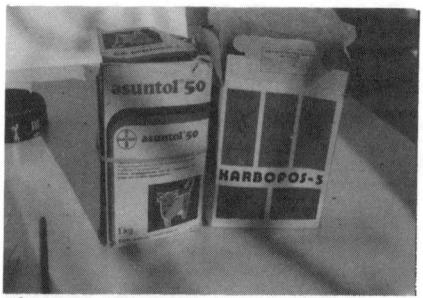

Fig 33

distributed by natural trophallaxis in amongst the bees in the colony. The product is based on the organophosphate insecticide Coumophos.

The acaricide must be freshly prepared, 1ml liquid in 49ml water. The emulsion is administered by a special dosing apparatus supplied with the chemical. Treatment is relatively easy and fast requiring only minutes per hive. The Perizin is trickled onto the bees between the frames in the brood chamber. Full strength colonies receive 50ml of the

Fig 34 *Applying Perizin to a swarm in W Germany*

freshly prepared emulsion. Weaker colonies and nuclei receive smaller doses (25ml for half sized colonies). Care must be taken to ensure that Perizin is only used when the ambient temperature is greater than 5°C. It was found that when the ambient temperatures were too low bees died from chilling after they were soaked. Perizin is generally used at the end of the season when the colony is low in brood. For control two treatments with a separation of seven days are necessary.

A short time after application the bees are licking up the Perizin from off each other. The bees ingest the medicine and pass it around amongst themselves through social interaction, food exchange etc. The chemical acts systematically on the mite which absorbs the medicament whilst it is feeding on the haemolymph of its host. Because of the way the chemical is distributed the first bees to take up the chemical receive a relatively high concentration, up to a toxic level, this is necessary so that the last bees receive sufficient medicine to effect the parasite.

Clive de Bruyn

The chemical is diluted as it passes from bee to bee. Obviously there is a narrow limit between a solution too strong which will kill bees (overdose) and one too weak to be effective. To reach an acaricide level which will be effective in killing mites, some initial bee mortality must be accepted. **Ref 5**

> **Ref 1** Levin. M.D., (1985) A new pest comes to Israel. Am Bee J. 125: 455-456.
> **Ref 2** Lubinevski. Y. et al (1987) Control of *Varroa jacobsoni* and Tropilaelaps clareae Mites using Mavrik™ in A. mellifera colonies under subtropical and tropical climates.
> **Ref 3** Liu, T.P. (1992) Fluvalinate and its After-effects Am Bee J 398
> **Ref 4** Morse R. (1991) Research review. Gleanings in Bee Culture, Sept 1991. pp 630
> **Ref 5** Infantidis. M., Thrasyvoulou. A., Pappas. N., (1988) Application of contact acaricide against varroa mites with contaminated proteinaceous food. Apidologie 19 (2): 131-138
> **Ref 6** Koeniger. N. & Fuchs. S., (1988) Control of *Varroa jacobsoni* in honeybee colonies containing sealed brood. Apidologie 19 (2): 117-130

SOFT CHEMICALS

These are the substances that many beekeepers would prefer to use instead of propriety hard chemicals. The fact still remains that these chemicals are not yet registered to treat colonies of honeybees in the UK. The legislation cares little whether a product is "natural" or man made. All medicines must be licensed. Anyone applying a product in contravention to the law is acting illegally.

Unless someone can find a way round the present regulations it does not seem as if any of the so called soft chemicals will be allowed here. It is not likely that anyone will apply for a licence and register such a product. A charge of £20,000 is often quoted for registration. In addition the application must be backed up with all the necessary research findings, test data and information to prove the products efficacy and safety. If, at the end of the exercise, the material can be obtained off the counter who is to profit?

If a latter-day Mr R.W. Frow was to come forward with "A new treatment for varroa" as the original did for acarine I am afraid we would not be allowed to use the product unless someone could come up with the licence fee and pay for the necessary data. Mr Frow unselfishly gave his "cure" to the nation. We all now live in a different commercial world I am afraid. On the other hand I am fearful when I hear of some of the substances which beekeepers have tried in an attempt to cure their colonies of varroasis. It may be that many of the colony losses which always accompany the introduction of varroa to a new area could be due to the treatments administered by well meaning beekeepers.

The New Varroa Handbook

Natural Plant Extracts

The use of plant derived "natural" oils against mites is well established. It is thought that these materials, many of which are strong smelling, disrupt the varroa from breeding in addition to killing them in some instances. Many different essential oils have been tested for their acaricidal properties and for their effect on bees. Unfortunately most of this work is anecdotal and not backed up with properly documented trials.

Dr Ritter of the Tierhygienisches Institute Freiburg, W. Germany found that oils of clove and wintergreen (methyl salicylate) were effective against varroa mites. 20% oil of wintergreen killed the mites without any significant effect on the bees. It was stated that the doses applied needed to be exact. The latitude between having a sufficient concentration to kill the mites and yet avoid damaging the bees was very small. There is no advice yet on how beekeepers can apply exact doses.

Professor Pickard of Cardiff University has reported that in the Philippines a "medicinal" plant is used to reduce infestation levels. This was referred to as Alagaw premna odorata (Verbenaceae) a tree growing up to 10m in height. The material is placed in the brood area separated from the bees by a wooden partition. He also mentioned that Snakeroot Eupatorium stacchadosmun, a spice and lemon grass Cymbopogon nardis are reported to be successful in controlling varroa in Vietnam. Both of these plants give off a strong aroma. All parts of the lemon grass give off a strong lemon aroma.

Dusts

Mites that are parasites on bees and wasps have well developed appendages to grasp onto their hosts. In order to escape the cleaning or grooming behaviour of its host a phoretic mite has to have a good method of holding on. It has been suggestion that the ambulacrum of the adult female varroa mites are not suckers or sticky pads but a structure with protractile claw like sclerites. These probably work as crochets that grasp the hairs of the bee, allowing the varroa to move rapidly on the bee and resist grooming. **Ref 1**

It has been noted by Sadov that mites have difficulty moving on wet or dusty surfaces **Ref 2** Apparently when the varroa's sticky tarsal pads are contaminated with dust the mite looses its hold on its host and drops off. The use of powders to control mite infestation in livestock is by no means unusual. Birds use dusting to maintain the condition of their feathers. It is also thought that the dusts act to control some of the external parasites and mites.

Dusts have been tried in France achieving success rates between 87% and 100%. The following substances have been used:
- Glucose powder
- Finely ground pollen
- Icing Sugar

- Pollen substitute
- Ground-up Eucalyptus (Casuarina)
- Talcum powder
- Chalk
- Flour
- Corn starch
- Diatomaceous earth
- Milk Powder
- Cellulose

The recommendation is to sprinkle about 50ml of the powder all over the bees active on the combs. Treatment is repeated at four day intervals. The treatment does not claim to kill the mites. An inlay protected by a screen would permit the collection of the living but temporarily helpless mites before they can remount a bee at a later stage. It would probably be better to use something sticky to trap the mites (see detection). The paper insert and mites can be burnt after recording the knockdown.

Six applications are said to give good control. Tins with holes, flour dredgers and baby powder tins are recommended as suitable applicators. The method is not patented and therefore can be tried by anyone.

> **Ref 1** Ramirez W.B. & Malavasi G.J., 1991. Confirmation of the Ambulacrum of *Varroa jacobsoni*: A grasping structure. Internat J. Acarol, Vol. 17, No 3. 169-172
>
> **Ref 2** Sadov, A.B., et al 1980. The pretarsus of the female varroa and mechanisms of its action. Veterinariya, Moscow, U.S.S.R. 36 -39

Formic Acid

Formic acid is a carboxylic fatty acid compound. with a formula of H-COOH. It is described as a "colourless fuming liquid with a pungent penetrating odour." It boils at 216°F and mixes well with water, alcohol, ether and glycerol. The acid is named after the formicine ants. who produce it in their poison glands and use it for:
1) chemical messages between individuals (pheromone)
2) defensive allomone
   (inter specific chemical messages)
3) defending the colony from intruders (formic acid is a powerful cytotoxin)

Formic acid has been extensively used in Europe since the late 1970s to control varroasis. As a naturally occurring acid rather than a man made pesticide it is favoured on the continent by many beekeepers. In the hive the acid breaks down relatively quickly and the possibility of bee products becoming contaminated by chemical residues is much reduced. I first heard about it from Beowolf Cooper 15 years ago. At that time the acid was applied in a bottle plus wick arrangement in the bottom of the hive. Claims for its efficacy vary from very good to only moderately successful. A great deal depends upon the concentration of the acid, the release mechanism used and where the acid is placed in the hive.

The New Varroa Handbook

Formic acid is able to diffuse into the brood cells where the varroa is generally protected from most chemicals. It is claimed that formic acid is capable of killing mites inside the brood. **Ref 1**. If this is the case it should even be effective in the active season where brood is present. Whilst formic acid is highly cytotoxic, it does not penetrate cuticle of insects easily. This could explain why formic acid does not cause heavy bee mortality when used in moderation.

Field tests have demonstrated that there is a risk to emerging and very young bees **Ref 2** Emerging adults bees may be susceptible due to their softer cuticle. Mites being smaller than bees are also damaged probably due to their thinner cuticle. It has been claimed that formic acid has been responsible for queen supersedure.

Effectiveness depends on:
- The infestation level
- The number of bees in the colony
- The space available for the bees
- The amount of brood present
- The ambient air temperature (From 12°C night to 25°C day)
- The ambient humidity
- Where the acid is applied.

The rate of evaporation is influenced by temperature. Other factors which influence evaporation would be the position of the colony and the air circulation around the hive. The acid is actually heavier than air but the action of the bees is sufficient to circulate the vapours throughout the colony. Present recommendations from Germany favour putting the chemical under the brood area rather than above. Less damage to the bees occurs when the bees are prevented from coming into contact with the acid.

Commercially prepared formic acid plates are available on the continent. **Ref 3**

A Briefing Note by the Veterinary Medicines Directorate

On Jan. 1st 1993 the single European market is due to be created and there has been much speculation about what this implies. In fact the single market will not materially affect the current rules on the supply of veterinary medicines in the UK.

Supply within the UK
Medicines marketed in the UK must have a UK product licence. A licence or other form of marketing authorisation issued in another common market state is not valid or recognised in the UK. This position will not change on January 1st 1993. It will remain an offence to market a veterinary product in the UK without a UK product licence.

Imports
Community law requires that anyone importing a veterinary medicinal product for sale or supply must have a valid UK product licence.
A licence will not be needed for the personal import of a pharmaceutical product for use by the importer to treat his or her own bees. However a licence will be necessary if the importer supplies the medicine to others e.g. within a co-operative organisation whether formal or informal.

Canadian DIY Practise

Material. 65% formic acid. Diluted from 85% commercial formic acid by adding water.

Method. 30 ml is applied onto 3 paper napkins placed on the top bars This is applied three times at weekly intervals resulting in killing 95% of the mites.

Bee reaction. There is immediate fanning and avoidance of the material. This reaction quietened down after 18 hours. No deleterious effects on the test colonies were observed.

German DIY Practise

Materials. 85% Formic Acid Two boards per hive 1.5 to 3mm cellulose or wood pulp. These should be capable of being placed on the hive floor with a 2" gap all round. Polythene bags to hold the boards. A plastic tray big enough to immerse the board in. A drawer system in the floor to hold the plates.

Method. Place the board in the tray. Cover with 25-40ml of the formic acid. Leave for the board to absorb the acid. When sufficient of the acid has been soaked up the board it is removed and placed in its plastic bag which is sealed. This is repeated for the second board. Two holes 1" diameter are made in the plastic bag.

Treatment. On a warm (15°C to 18°C) evening when bees have stopped flying the two boards are placed in the drawer of the varroa floor with the two holes facing down towards the floor. Boards should be raised on wooden laths to allow a free circulation of air around the plates. Bees must not be allowed to come into contact with the acid. The acid will evaporate and destroy mites. The hive entrance should be fully open whilst the plates are in. The minimum treatment time is two days. The treatment is repeated four more times at intervals of 7 days. It is claimed that 95% of the mites present in an infested colony can be eliminated with this treatment.

Precautions. The impregnation should only be carried out in a cool well ventilated place, preferably outside. Rubber gloves should be used. The acid should not be allowed to come into contact with any metal parts. If the concentration of acid used is too strong the bees may be driven out of the hive. If this happens remove the boards.

After treatment the formic acid content of the honey can increase to 1000mg/kg. Formic acid is very volatile and therefore its residue in honey rapidly decreases with time. Nearly all honey contains a small fraction of natural formic acid along with other acids such as gluconic and acetic acids **Ref 4** A very high content of formic acid is found in honeydew honey **Ref 5**. It has been found to vary from 1000mg/kg in honeydew honey to 626 mg/kg in chestnut honey. The formic acid content of honey from non-treated colonies can range from 70 to 85 ppm **Ref 6**.

A normal level for most honeys is about 20 mg/kg (7). In experimental colonies honey samples collected two weeks after treatment with formic acid contained 561 ppm and 64 ppm when measured one month later. Formic acid can get into beeswax easily but it quickly vaporises and it does not build up in the wax. **Ref 8**

Formic acid is cytotoxic **Ref 9** and it must be handled with care. There are life threatening consequences which can arise form misuse. Organic acids are dangerous in their concentrated form. Concentrated formic acid can cause painful burns on the skin. If you spill formic acid on your skin it will usually penetrate the full thickness of the skin causing staining and scar tissue which takes some time to heal. If accidentally imbibed it would cause severe burning of the mouth and throat with subsequent vomiting.

Fumes in an enclosed space are dangerous to life and health. The permissible exposure limit for formic acid in air is 5 parts per million averaged over an 8 hour work shift **Ref 10**

When formic acid was applied to control acarine it was stated that the honeybees did not show any observable ill effects of the formic acid at the levels applied. (three applications of 22, 44 and 66 g/colony of 65% formic acid one week intervals, respectively, by pouring onto a cheese cloth plate placed in each colony) **Ref 11**. On the other hand reports from Germany suggest that many colonies treated with formic acid in 1990 did not do very well. **Ref 12**

Ref 1   Fries. I., (1991) Treatment of sealed honeybee brood with formic acid for control of *Varroa jacobsoni*. American Bee Journal May: 313-314.
Ref 2   Liebig. C. (1985) Uber Spatwirkungen der Ameisensaure-Behandlung. Report. Arbeitsgemeinschaft der Deutschen Bienenin stitut. Bonn 1985. Apidologie 16: 191
Ref 3   IMP plates Illertissener Milbenplatte registered Germany
Ref 4   White. J. W. (1975) The hive and the honeybee. ed Dadant & sons.
Ref 5   Wachendorfer. G., & Keding. H., (1988) Evaluation of residues in honey after the use of varroa treatment chemicals in view of the official food control. Allgeneine Deutsche ImKerzeitung 22: 414-421.
Ref 6   Clark. K., J., & Gates. J., (1982) Investigations of the use of formic acid for the control of honey bee tracheal mites in British Colombia. Proceedings of the annual convention of Canada Honey Council (in press)
Ref 7   Stoya, W., et al (1986) Ameisensaure als Therapeutikum gagen Varroatose und ihre Auswirkungen auf den Honig. Deutsche Lebensmittel-Rundschau. 82: 217-221
Ref 8   Liu. T.P., (1992) Formic Acid Residue in Honey and Beeswax. American Bee Journal. Oct: 665

Ref 9 Blum. M.S., (1978) Biochemical defences of insects. Biochemistry of insects ed. Morris Rockstein. Academic Press, New York Liesivuori. J., & Savolainen. H., (1991) Methanol and Formic acid toxicity. Biochemical mechanisms. Pharmacology and Toxicology. 69: 157-163.
Ref 10 Clark, K 1992 Human exposure to Formic Acid from Applications for the control of Honey bee Trachael mites. Pesticides Directorate, Agriculture Canada. In support of CAPA proposal
Ref 11 Liu. T.P. & Nasr. M., (1992) Effects of Formic acid treatment on the infestation of Tracheal mites, *Acarapis woodi*, in the Honey bee *Apis mellifera*. American Bee Journal Oct: 666-668
Ref 12 Koeniger, N., (1990) Situation und Persperktiven der Varroatosebehandlung Deutsches Imker-Journal 3/90

Lactic Acid

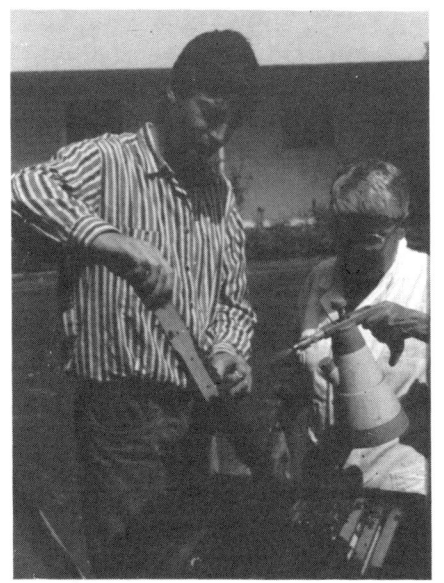

Fig 34b *Applying Lactic Acid. Kirchain, W. Germany*

Lactic acid is an organic acid It is present naturally in milk, hence its name, and meat. It is a natural constituent found in honey as well as yoghurt, wine and cheese. Residues of the acid are considered harmless. It is relatively easy and cheap to obtain.

15% lactic acid solution can be applied by a hand sprayer directly to bees on the broodless combs. Each comb is taken out separately and the acid sprayed on in such a way that all the bees are wetted by the spray. Three applications to brood free colonies in winter has been found to be effective in removing 90% of the mites in a colony.

The method is labour intensive. It is also only effective if used when there is no brood in the colony. This limits its use to the short broodless period in winter or to artificial swarms in the active season. Soaking the bees, especially in winter, can put them under considerable stress. The ambient temperature should not be too low as to chill the bees and cause them to fall from the cluster.

Lactic acid is corrosive capable of causing skin burns and the operator must be suitably protected with chemical proof gloves.

3.8 BIO-TECHNICAL CONTROL

A husbandry technique used to limit the build up of varroa mites in a colony can be referred to as a "Bio-technical Method". Bio-technical methods generally mean colony manipulations which take advantage of the natural behavioural patterns of both honeybees and varroa mites. No bio-technical method of control can keep the mite at bay continuously entirely on its own. Occasional chemical treatment might still be required.

The aim of any management must be to keep varroa populations low enough so that the honeybee colony can live with the infestation. The possibility of eradicating the disease is not realistic.

Advantages

1) No residues
2) No danger of varroa becoming resistant. However it must be borne in mind that the mite will adapt to management techniques as well as to chemical attack (Drone trapping.)

Difficulties

1) Labour intensive
2) A high degree of competence is necessary on the beekeeper's part.

By restricting the brood frames available for the mites to breed in the number of mites raised can be limited. Many variations exist depending on local conditions. Any method that helped to remove a proportion of breeding female mites from the colony must help.

COMB TRAPPING
Caging the Queen

This method was developed by Dr Maul (W. German Kirchhain Beekeeping Research Institute)

The original method relies on caging the queen on comb for about four weeks (separate periods of 9 days on three different empty drawn comb) during the brood rearing season. The varroa mites, relying as they do on brood for reproduction, have to migrate to the larvae in the cage. When the brood is capped the frame is removed and the mites destroyed. It is claimed that this technique is capable of removing about 95% of the mites. Having witnessed this technique in Germany I must report that trapping the queen permanently is important. In one of the trials prepared for us it was most embarrassing for our hosts to discover that the queen had escaped. However it taught us all that attention to detail is important.

Timing is critical. The method restricts normal colony development. If applied too early it will effect the honey flow. Applied too late and it will reduce the

Clive de Bruyn

Empty brood frame inside a queen excluder screen.

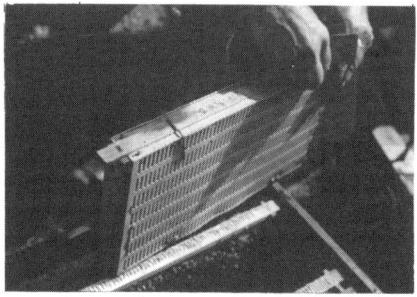

The queen is free to move about within the cage and lay eggs.

colony's ability to build up for winter. What may be suitable for one area may not necessarily work elsewhere. The length of the active season and the onset of the honey flows need to be considered. Some people could even discover that it might be useful in controlling swarming if applied at the correct time. A restriction of brood rearing in the middle of a good flow

Worker bees and varroa can come and go as they please.

may result in an increase in your honey crop. Many more bees will be available for foraging rather than feeding brood. The method needs to be tried and assessed in the light of one's own personal circumstances and local conditions.

Day 1 Find queen, place her on empty drawn comb close up the excluder cage trapping the queen. Place the cage in the centre of the brood nest.

Day 9 Remove the comb of brood where the queen was trapped and return it to the hive. Mark it with the date Comb A.
The queen is again trapped in the excluder frame with another empty drawn comb for a further period.

Day 18 Remove the comb of brood where the queen was trapped and return it to the hive. Mark it with the date Comb B.
The queen is again trapped with another empty drawn comb for a further period. The comb of sealed brood Comb A is removed and the mites destroyed.

135

Day 27 Remove the comb of brood where the queen was trapped and release her into the brood nest of the colony. Mark the comb where the queen was trapped with the date Comb C. The comb of sealed brood Comb B is removed and the mites destroyed.

Day 36 The comb of sealed brood Comb C is removed and the mites destroyed.

The control is best carried out in June/July (Germany) when the colony can afford the loss of brood. Three frames of brood = 15 000 bees.

COMB TRAPPING
In broodless nuclei

A brood frame of uncapped brood is inserted into a nucleus that only has sealed brood. The comb of unsealed brood acts as a magnet for the free varroa mites not involved in breeding. The comb and the mites can subsequently be removed.

Artificial Swarm (1)

Developed by Walter Gotz of the Institute fur Bienkunde, Oberusel.

1) Build up the colony until there is brood in two brood boxes. The colony is allowed to store its honey in supers over a queen excluder.
2) On day 1 of the treatment a second queen excluder is inserted between the two brood boxes to trap the queen in the top box.
3) On day 9 the upper brood chamber and the queen is placed on another stand in the same apiary. The queenless colony, reinforced by all the foraging bees has a queen cell introduced. Acceptance.
4) On day 18 two combs containing a large proportion of unsealed brood are removed from the queen right brood box and placed in the colony with the virgin queen which must have now emerged. These combs will act as varroa traps as they contain the only brood available for the female mites to reproduce in. The mites will find this open brood extremely attractive. At this stage the stock with the old queen is moved again to another part of the apiary.
5) On day 27 the two trap combs are removed and the mites destroyed. The colonies can now be left to build up independently or united after killing the old queen.

The technique may enable queen rearing and swarm control measures to be integrated with varroa control.

Artificial Swarm (2)

The "Tubingen Technique" developed by Professor Engles. Following the early flow the bees from strong colonies are shaken into swarm boxes so that the adult bees could be treated chemically. Each unit was given a new queen and allowed to build up (fed as necessary).

COMB TRAPPING
Use of Drone Brood

Because of the preference of mites for reproducing in drone cells, systematic removal of drone brood before the drones emerge can reduce the number of mite in the colony. The method is only effective at low levels of infestation (up to 500 mites per colony.)

The varroa mite has to feed on the haemolymph of the brood before it can lay eggs. The hormone necessary to trigger egg laying occurs in greater concentrations in drone larvae than worker larvae and particularly in the case of *A. cerana*.

There is a hope that a pheromone produced by drone larvae can be isolated **Ref 1**. It could be that the varroa mites are attracted by this pheromone. Such findings hold promise for the development of a trap which can be impregnated with the pheromone. This trap could be left in the hive to capture mites. Until such time we shall have to resort to using drone brood.

Removal of sealed Drone Combs

Drone comb can be placed in the hive the previous Autumn where it will be occupied early in the year. Early removal of drone comb in May might not be the best technique. German experience has shown that infestation levels in these combs were low at this early period of the year. We shall have to verify these findings in the British Isles. During the active season each colony receives frames with drawn drone cells at the edge of the brood nest.

The capped drone comb is removed regularly and replaced with drawn drone comb or drone foundation. The drone comb must be removed before the drones emerge The number of mites in 100 drone cells should be checked so that some attempt can be made to monitor the mite levels and obtain data on the effectiveness of the treatment. Although I do not believe that uncapping drone brood is a very reliable method of detecting low levels of mites I think that examining a set number of cells at regular intervals could be useful.

Drone comb which the beekeeper

Fig 38 *Cutting out drone brood for mite inspection*

Fig 39 *Separating the pupa from the mites for counting.*

wants to use again can be decappped and the brood knocked out or washed out with a jet of water. Immerse the comb in hot water at 55°C for thirty minutes. The temperature control is critical as the wax will begin to soften at 62°C.

From my experiences gained in 20 years of raising queens it is just not possible to put drone foundation or drawn drone comb into a colony and automatically get slabs of drone brood. Is it just that my bees are awkward? I know that if I leave a gap in the brood nest, say by putting a shallow frame in a deep box, then there is a good chance that the bees will draw out some drone comb and the queen will lay in it. This would probably be better than a solid deep frame of drone brood anyway. It would seem unnatural in nature for the queen to lay so many drone eggs in only a short time. Probably the colony requires a constant supply of drones continuously through the season. If this is the case then the colony might be more inclined to co-operate if the beekeeper takes measures to ensure that there is a constant supply of drone cells. What is required is an opportunity to continuously trap varroa mites throughout the breeding season. It would be better to use many smaller portions of drone brood regularly than one large area infrequently. If the sealed drone brood is constantly removed there will be a pressure on the colony to produce more. However knowing that bees do not use the same reasoning as I do, I will wait and see.

I noted in Germany that they had in operation a deep frame split vertically into three sections of drone comb. Each piece of drone comb could be removed separately. If on each visit one piece of sealed drone comb could be removed and replaced with a section of drawn empty drone comb then the queen and the bees might co-operate in providing a constant area of uncapped drone brood for the mites to enter for breeding. If one is working on a 9 day inspection system then the rotation would be one of 27 days, see table.

Day 1: one third empty drawn drone comb inserted (no 1 in)
Day 10: one third empty drawn drone comb inserted (no 2 in)
Day 19: one third empty drawn drone comb inserted (no 3 in no1 out)
Day 28: one third empty drawn drone comb inserted (no 1 back in no 2 out). And so on.

If the first section were placed in the hive in May 1st and the exchange maintained until August (colony permitting) there would be constant supply of cells available throughout the season. Quite a few varroa mites could be trapped. The co-operation of the varroa mites is also necessary. In my experience I have not always found that the varroa mites are present to a greater extent in drone brood. I recall that the colonies I inspected on the Isle of Wight last year had significantly more varroa mites in the worker brood. Once again more experience is necessary.

Clive de Bruyn

A more simple way would be to insert a shallow frame in and remove the sealed drone comb by cutting it off. In this case one would merely return the "worker" brood in the shallow back to the colony for the bees to draw out some more drone comb. If one wanted to withdraw the drone comb regularly three shallow frames would be required. Some experience will be necessary to find the best place to put the drone cells to ensure the queen lays in them. This might alter through the season.

It should be obvious that the effectiveness of this technique will be negated if there is a high proportion of drone comb elsewhere in the brood nest. Indeed the presence of any free drone cells in the colony will allow some breeding, the more drone brood the less the chances of removing mites. Although of course the varroa will be able to reproduce, albeit at a slower rate in worker brood.

As a practising queen rearer with some desire to exercise an influence on the drones my queens mate with how am I to proceed? My drones are my most valuable product. Special techniques will be needed in drone rearing colonies. Probably these will have to be treated constantly with an acaricide to keep mite numbers down and allow drones to be reared.

Queen Removal

During the main swarming period (May to July) a brood comb of bees and the queen can be removed. The colonies were allowed to build queen cells but the virgin was not allowed to mate. At the end of three weeks the colony can be requeened. It should be possible to integrate such a system with judicious use of an acaricide when the brood is low.

Ref 1   Le Conte, Y. et al (1989) Attraction of the parasitic mite varroa to the drone brood larvae of honey bees by simple aliphatic esters. Science 245: 638-639.

3.9 INTEGRATED CONTROL

Integrated pest control requires an informed beekeeping population. This could be achieved through an effective extension service able to communicate with beekeepers. These extension officers should be backed up with suitable educational material etc. To allow all beekeepers to understand and put into practice a more effective management programme for controlling varroasis. With the current cut back in non vocational education and the loss of County Beekeeping Instructors beekeeping education is at a serious disadvantage.

The FAO defines integrated pest control as a management system that, in the context of the associated environment and the population dynamics of the pest species, utilises all suitable techniques and methods in as compatible a manner as possible and maintains the pest populations at levels below those causing economic injury.

If we look at what has been achieved in Germany it can give us some hope that integrated control would also be possible here. The following is an example of an integrated control for Varroasis as practised in Germany.

The technique relies heavily on biotechnical control and does not use any hard chemicals. The methods have been applied to 45 honey production colonies over a period of seven years (1982-1989). The main pointers of the system are:

1) Diagnosis
2) Removal of drone comb
3) Use of comb trapping techniques
4) Treatment with formic acid
5) Use of shook-swarms.

1) Diagnosis
One needs to know the level of infestation in a colony at various times in the year. A variety of techniques need to be used depending on circumstances. The same method is not necessarily applicable throughout the season. The armoury of techniques include :
- Examination of winter debris
- Use of summer inserts
- Occasional examination of worker brood
- Examination of drone brood
- Level of mites in trap combs
- Dead mites collected on inserts after chemical treatment
- Physical appearance of the bees.

An appropriate response can be formulated after the level of infestation has been estimated.

Threshold levels have been worked out in Germany. These may not apply in the UK. Work recently undertaken by the National Beekeeping Unit is to look into the various detection methods currently available and relate these to the mites present in a colony. Until this work is finished we will have to rely on the findings of others.

Infestation level 1:
0-500 mites   Far below damage threshold, colony under no threat

Infestation level 2:
500-1500 mites   Still below damage threshold, colony may be slightly impaired

Infestation level 3:
1500-3000 mites   Damage threshold reached,   colony under threat

Infestation level 4:
over 3000 mites   Damage threshold decidedly exceeded,
                                      colony under stress,
                                      collapse likely.

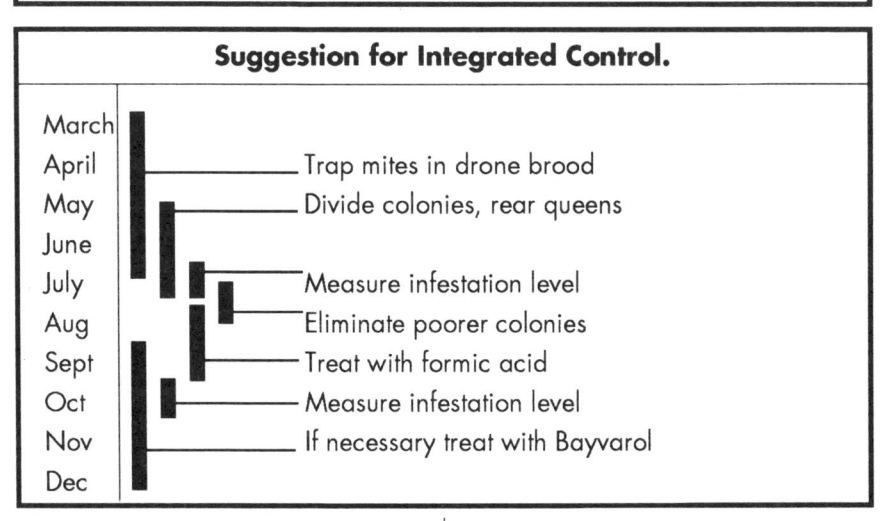

Suggestion for Integrated Control.

March	
April	Trap mites in drone brood
May	Divide colonies, rear queens
June	
July	Measure infestation level
Aug	Eliminate poorer colonies
Sept	Treat with formic acid
Oct	Measure infestation level
Nov	If necessary treat with Bayvarol
Dec	

German experience shows that every female mite that dies naturally each day during the beekeeping season can be taken to represent 120 live mites in the colony. During early spring into summer mite build transfer on bees is low as most of the females are breeding in brood combs. In Spring 1 mite on the floor = 500 in the colony In Summer 1 mite on the floor = 150 in the colony.

Control methods

Experience is needed to interpret the diagnosis. It is also important to work with the bees rather than to implement a strict control methodology.

Success of any treatment will depend on the length of the "brood free period" which will depend on the severity of the

The New Varroa Handbook

winter and the type of bee. The build up in spring Condition duration and timing of the flows The natural colony cycle:
Swarming, the brood area, size and growth.
- The onset of winter
- The age of Queens
- Stress factors
- Other diseases
- Poisoning
- Management methods of the beekeeper
- Level of infestation and development cycle
- Infestation levels in other colonies in the area.

In a period of seven years these beekeepers have demonstrated that it is possible to control varroa mites within reasonable levels with only the minimum use of chemicals. This has only been possible because of the time and trouble the beekeepers have taken to monitor the number of varroa mites present in their colonies. Armed with this knowledge they have been able to take appropriate action which were in sympathy with the prevailing conditions. Their experiences should provide us all with an indication of the way forward. **Ref 1** Examination of annual figures kept by beekeepers have shown that the build up of varroa populations can be maintained below the damage threshold by means of skilful bee management combined with the limited use of medicaments.

The use of biotechnical control in the early years of the infestation can delay the build up of mites in a colony. This will increase the period before chemicals have to be resorted to. If natural selection is to be given a chance then a period must be allowed for the bees to build up a resistance. Too early a use of chemicals will slow down the process of natural selection. It will allow susceptible strains of bee to survive instead of being eliminated.

Ref 1 Etteler, H., Henkemeyer, J. (1991) Varroa bekampfung. Deutschs Imker-Journal 6/90: 236, 253-254

Clive de Bruyn

TREATMENTS (Non Chemical) 3.10

There are no absolute methods of controlling varroa either chemical or biological. Various methods have been advocated that cannot be strictly defined as boi-technical or chemical. I include them here merely because they are recorded in the literature. I have not been able to find anyone who could give me any first hand account of their efficacy.

Dusts

It was noted by Sadov et al that mites have difficulty moving on wet or dusty surfaces **Ref 1**. The use of powders to control mite infestation in livestock is by no means unusual. Birds use dusting to maintain the condition of their feathers. It is also thought that the dusts act to control some of their external parasites and mites.

Mites that are phoretic on bees, wasps and other arthropods have well developed tarsal claws to grasp onto their hosts. In order to escape the cleaning or grooming behaviour of its host, a phoretic or parasitic mite has to have a good means of clinging or holding onto its host whilst still allowing for swift movement. Varroa survives for long intervals on adult bees in the absence of brood.

It is a suggestion that the ambulacrum of the adult female mites are not suckers or sticky pads but a structure with protractile claw-like sclerites. These probably work as crochets that grasp the hairs of the bee, allowing the varroa to move rapidly on the bee and resist grooming. **Ref 1, 2**

Prof. W.B. Ramirez, Head of Entomology in the Phytotechnic Department of the University of Costa Reca recommends the use of dusts to control mites. Apparently when the varroa's pads are contaminated with dust the mite loses its hold on its host and drops off.

Dusts have been tried in France achieving success rates between 87% and 100%.
Powders used:
- Glucose powder
- Natural substance in honey
- Finely ground pollen
- Icing Sugar
- Innocuous Pollen substitute
- Ground up Eucalyptus (Casuarina)
- Australian spruce
- Talcum powder
- Chalk
- Flour
- Corn starch
- Diatomaceous earth
- Fullers Earth
- Milk Powder

The recommendation is to sprinkle about 50ml of the dust all over the bees which are active on the combs.

Treatment is repeated at four day intervals. Six applications are said to give good control. Tins with holes, flour dredgers and baby powder tins have all been found to be suitable applicators. The mites are not killed by the dust. After the mites fall off they can try to find another host. An inlay protected by a suitable screen would permit the collection of these temporarily helpless mites for disposal before they can remount a bee. The paper insert and mites can be disposed of by burning.

The method is not patented and therefore can be tried by anyone. It does not introduce harmful chemicals.

Special Plastic Comb

Apis Nova Products P.O. B 545, D-7320 Goppingen, W. Germany.

This comb has cells where the base is larger than the mouth. These conical cells have thickened walls towards the mouth so that the cell walls are not parallel but slightly conical. This results in a reduction of 40% in the total number of cells per unit area. Compared with normal comb however, each cell is larger. The two sheets of bottomless cells are pegged together sandwiching a plain plastic sheet which acts as the septum.

The theory is that larger cells mean more food for the larva and more food causes faster development. Dr Mautz of the Bayerische Landeranstalt fur Bienzucht. W. Germany has conducted trials with this comb. Artificial swarms were given the plastic comb and fed. The combs were accepted and brood was produced. The experiments concluded that the sealed brood phase was shortened on average by one day. This is attributed to the extra food that each larva receives due to the larger cell floor.

The premise is that the shorter capping period will reduce the number of female mites able to mature and mate before the imago emerges. This should significantly reduce the rate at which mite populations increase. The theory seems to be born out in limited trials carried out in the Black Forest, reported in the ABJ. **Ref 3**

Thermal Treatment

Varroa mites do not appear to be able to stand high temperatures as well as the bees can. This has been used by some as the basis of a thermal treatment to rid bees of varroa mites.

In laboratory experiments it has been observed that varroa mites will release their hold on their host when subjected to temperatures in excess of 40°C. The Russians have developed this into a technique for controlling varroasis. Treatment is recommended in the late autumn. Treatment too early can promote additional brood rearing. The ambient temperature should be between 0°C and 10°C.

Infested bees are removed from the comb. The bees are shaken or blown into cylindrical wire mesh containers. These containers are placed in the treatment chamber. Inside the chamber the mesh containers can be revolved on a horizontal axis. It is also important to shake the cage from time to time to disperse the bees and prevent them clustering. Dry hot air is blown into the chamber (47°C with a R.H less than 20%). Conditions are carefully controlled. The mites soon start to fall off the bees within 12 to 15 minutes. Treatment is continued until the mites stop falling from the bees. The temperature is lowered to 35°C and the bees are then shaken back into the hive. It is claimed that at least 95% of the mites present on the bees can be removed by this method. One treatment a year seems to suffice. **Ref 4**

The method is fairly labour intensive (each colony takes about 30 minutes). Unfortunately workers outside Russia have not found this to be an efficient method of removing mites. Workers in Germany failed to confirm the Russian results. **Ref 5** In Holland a technique was developed for treating an entire colony in its hive. The hives were held in a room heated to 50°C. The floors were removed from the hives at various intervals (18, 45 and 60 minutes). The hives were then removed from the chamber. A follow up treatment with Perizin revealed the presence of hundreds of mites whereas not one mite dropped off during the heat treatment. Heating the entire colony would thus appear to be ineffectual in removing mites. A simple method for treating entire colonies will probably never be effective as the bees in the colony are able to control the temperature below that at which mites will fall off. **Ref 6**

Ref 1   Sadov, A.B., et al 1980. The pretarsus of the female Varroa and mechanisms of its action. Veterinariya, Moscow, U.S.S.R. 36 -39
Ref 2   Ramirez W.B. & Malavasi G.J., 1991. Confirmation of the Ambulacrum of *Varroa jacobsoni*: A grasping structure. Internat J. A carol, Vol. 17, No 3. 169-172
Ref 3   Von Posern. H., (1988) The synthetic comb, a new weapon to fight the varroa mite
Ref 4   Komissar, A.D. (1985) Heat Treatment of varroa infested honeybee colonies. Apiacta 20 (4): 1143-117
Ref 5   Hoppe, H. & Ritter, W. (1986) The possibilities and limits of Thermal treatment as a Biological Method of Fighting Varroa . Apodologie 17 (4): 374-376
Ref 6   De Ruijter, A., Van den Eijnde, J., Van der Steen. J (1988) Research report of the Research Centre for insect pollination and Beekeeping. Ambrosiushoeve, Hilvarenbeek. Netherlands

3.11 BREEDING RESISTANCE

Over the many millions of years that bees have been around they have been exposed to attack from a wide range of enemies. Honeybees possess a great number of mechanisms that protect them against pests parasites and predators ranging from microscopic viruses to large mammals. Arguably the most harmful species they have come up against so far has been *Homo sapiens*. In a cynical mood I sometimes think that if only a way could be found to eliminate man without harming the honeybee then the world and the environment would be a better place.

Bees have many methods of resisting diseases and infestations. A colony's survival will depend on the degree to which these specialist properties are developed. If there were no human interference then in time a balance could be struck between a host and its parasite. *Varroa jacobsoni* is totally dependent upon its host for its survival. It is not in a highly specialised parasite's interest to have too devastating an effect on its host. The mites which are sufficiently damaging to kill their host would soon commit themselves to extinction.

Within the varroa species there will be differences. Some mites may be more benign. These will cause minimal damage to the host. Bees with some degree of resistance will survive. These genetic traits will eventually spread throughout the honeybee and mite populations. The result will be the eventual establishment of varroasis as another endemic pest which will be tolerated in most colonies. Where additional stresses are present some colonies will exhibit damage and occasionally succumb. The majority will survive despite the mites presence. This seems to be occurring already in some instances. What I do not know is how long it will be before this state of affairs is attained here.

Nature will not be allowed to run its course in the United Kingdom. Initially the control methodology will be one of management and medicaments. Economic survival may dictate that "One cannot afford to mess about looking for natural resistance when there is a honey crop to get." On the other hand the search for bees able to tolerate varroa could be speeded up if "someone would be prepared to put in the few £1,000,000s necessary". Some chance! My hope rests with the survival record of the bees themselves. They will survive varroasis in spite of *Homo sapiens*. Whilst colonies in the hands of beekeepers are being treated, be it biotechnical or chemically, the opportunity for natural resistance to manifest itself will be interfered with. Feral colonies which are not subject to such treatment could be worth watching.

A vast amount of data exists on the adaptability of The Western Honeybee to new environments. *Apis mellifera* shows much variation in the huge territory it now inhabits. There is great variability between individuals and colony behaviour. We are aware that there are differences in colour, size, tongue length and other morphological aspects. The colony's prolificacy, aggressiveness, hormones, behaviour and communication are also of significance for the colony's fitness to survive.

Apis cerana

As has already been stated, varroa is apparently not found to be a serious pest of *Apis cerana*, its original host. The exact mechanisms by which the cerana bees are resistant to the ravages of varroa are only poorly understood. It is evident that *A. cerana* colonies are able to survive and prosper in spite of mite infestations.

It is known that female mites reproduce in the drone brood of cerana not the worker brood. Because the mite is only able to reproduce satisfactorily in the cerana drone brood the mite population is not allowed to build up so easily as it can in mellifera where both drone brood and worker brood are suitable for breeding. It has been argued that the perforated drone capping plays a part in allowing the varroa mite to become a specialist parasite on *Apis cerana* **Ref 1**. If these conclusions are correct it would explain why the varroa mite restricts its brood rearing specifically to drone brood in its original host *A. cerana*. In addition the cerana workers are able to detect worker brood infested with varroa. This hygienic behaviour is related possibly to the release of a brood pheromone.

The metamorphosis from pupa to imago is the result of hormone changes that take place relative to feeding, temperature and genetic differences. It has been found that there are differences in the concentration of juvenile hormone in the Apis species and this is suggested as a possible reason why *V. jacobsoni* prefers to lay eggs in *A. mellifera* worker cells and not in *A. cerana* worker cells **Ref 2**.

There are also behavioural differences. *A. cerana* workers are able to detect varroa mites and remove them from their bodies and from brood cells. The comparatively large mites can be gripped by the bees with their mandibles. Apparently the mites are "bitten" and damaged. Many mites are killed in this way. The damaged mites can be found on the floor of the nest. This grooming behaviour manifests itself in many ways:
- self-grooming
- nest-mate grooming
- group cleaning behaviour
- grooming dances.

It is said that such behaviour contributes significantly in reducing mite populations in the hive.

Africanized Bees

Africanized bees in the tropical areas

of South America seem to be able to live with varroa without any noticeable symptoms. These bees are derived from descendants of the *Apis mellifera scutellata* introduced to Brazil in the 1955. The so called "K___ Bees".

In 1989 I was able to search for varroa in more than a 100 colonies at numerous widely separated locations. The colonies belonged to commercial beekeepers, amateurs, teaching apiaries and research centres. The colonies were situated at the coast and in two major cities as well as in the forests and orange groves. Varroa mites could be found in every colony where they were looked for. All the beekeepers I met during my stay were able to live with varroa without the necessity of applying medicaments. The mite populations never seemed to rise to damaging levels. No damaged bees or other symptoms were ever seen. I met no one who treated for varroa. In fact there are no chemicals registered for varroasis in Brazil.

A natural limit seems to be placed on the varroa rate of reproduction. in these bees. The exact mechanism responsible is not known. It may be possible that the varroa mites in Brazil are descended from a more "benign" strain. We do know that the bees have a shorter capping period than our European mellifera bees. This must reduce the time that the mites have available to them for reproduction. African bees seem to have a high reaction to disturbances. This might manifest itself in an increase in grooming behaviour. Local conditions and the climate may play some part **Ref 3**. Varroa does not seem to be a problem in the tropical lowlands but some colony damage occurs in colonies at a high altitude or in the more temperate south.

Some workers have gone on record that varroa cannot be a serious problem in the tropics because of climatic influences. However, *A. mellifera* bees introduced into Burma, Thailand and Malaysia soon become heavily infested. These bees cannot survive without extensive management controls on the part of the beekeeper. This would indicate that climate alone is not responsible for the amount of damage caused. Colonies of mellifera bees had been treated with acaricide when I visited Thailand in 1992. Adjacent cerana stocks needed no treatment. I was told that mellifera could not survive without treatment.

Worker bees in African races develop in 19 to 20 days **Ref 4** with a post capping period of 11.5 to 20 days in contrast to European races which are supposed to take 21 days. This relatively short period of capping could limit the reproduction rate of varroa. This may be one of the genetic factors which leads to low population levels of varroasis in the tropical areas of South America. There are known differences in the Juvenile Hormone of bees from Africa. It has been proposed that the higher levels of JH in Africanized bees causes them to age physiologically faster than European bees. Certainly it is most noticeable that they seem to be faster at everything than their European cousins. They develop faster, they move quicker on the comb and they live shorter lives.

Work in Uruguay has established that 70 to 90% of the female varroa mites in worker brood produce no offspring. Reproduction seemed to take place principally in the drone brood.

Experiments

In 1988 Africanized bees from Venezuela were imported to Oberusel (near Frankfurt W. Germany) **Ref 5**. A free flying experiment was carried out with the progeny of these bees in the winter months. The natural varroa mite mortality was measured and compared with two control groups.

1) Pure Africanized
2) Pure Carniolan
3) Africanized x Carniolan

Natural mite mortality was monitored using paper inserts on the hive floors. These were inspected weekly and the number of mites counted. Brood was also examined where it was present. At the end of the trial the surviving colonies were killed with sulphur and the total mite population counted.

Normally about 10% of the female mites present in an infested colony die and fall to the floor in winter. In the experiment the following results were obtained:

Carniolan   14%
Africanized, Hybrid   34%

Such levels of mite mortality have not been recorded in Europe before.

This winter mite mortality could be significant if one remembers that the mite population present at the beginning of the season can have a dramatic influence on the population dynamics and build-up during the brood rearing period.

Before anyone considers that the African bee is the answer I must point out that these bees have a reputation of extreme aggressiveness plus the fact that the Africanized colonies suffered in the winter. Only 2 of the 9 colonies attained a total bee weight of 500gm by the end of the experiment. All the Carniolan exceeded 500gm.

Bees with shorter capping periods have a natural biological advantage that works by reducing the number of mature fertile female mites that are produced during each cycle. One must remember of course that mites will still be able to breed in the drone brood which will still take longer to emerge. It has been demonstrated that the duration of the post capping stage in Apis is a highly heritable characteristic. **Ref 6**. Therefore, the selection of bees having a shorter development cycle may be one mechanism worth studying.

12 days   for A.m. ligustica
11.2 days   for Africanized
9.6 days   for A.m. capensis

The period for development of brood is influenced by genetics, temperature and feeding. Worker brood is always quoted as taking 21 days from the laying of the egg to the emerging of the imago. However, this can be up to 1 - 2 days longer on the edges of the brood nest, due to poorer feeding and lower brood temperatures which can occur in these areas. Conversely worker brood at the centre of the nest may develop in less than 21 days.

There is evidence that there is a variance between the capping periods of various European races of 9 hours between strains and 19 hours between individual colonies. **Ref 7**. Seasonal factors are also significant. The average capped period was found to be seven hours shorter in the early part of the year compared to late summer. The effect of only a few hours less in the capped period is surprisingly high and stresses the importance of this factor in selecting for varroa tolerance.

I have seen a report of work in Europe where it was found that colonies that produced queens in a shorter period (14 days instead of 16) also produced worker bees in a shorter period **Ref 8**. This may be a good way to screen likely colonies for the genetic ability to shorten the brood cycle.

The Cape honey bee from the southern tip of South Africa, *Apis mellifera capensis* has a shorter brood capping period than the European races of honeybee. Theoretically only 21% of the mites will be able to reproduce. Mathematical models are apt to fall down in the light of practical field experiments so they should be used with discretion.

Research workers in Germany have carried out experiments with capensis colonies. **Ref 9**. Although only 10 colonies were used (5 carniolan, 5 capensis) in this research the results are in agreement with the theory. Colonies were monitored for mite infestation. Dead mites were collected weekly on screened bottom boards In addition mite counts were made on adult workers collected from the brood area and 100 brood cells at regular intervals. The researchers found that the *A.m. capensis* colonies appeared more tolerant to varroa than the *A.m. carnica* bees. It appeared that the build up of the mites in the capensis stocks had been restricted as had been predicted with the mathematical models.

The number of mites present on bees and in the brood in October were compared and the carnica colonies were reckoned to be three times more infested than the capensis. This winter mite mortality could be significant if one remembers that the mite population present at the beginning of the season can have a dramatic influence on the population dynamics and build-up during the brood rearing period. The Cape bees were also shown to be more efficient in mite removal behaviour than the carniolan bees. This grooming behaviour observed in these African bees may be an additional genetic factor.

In experiments where mites were introduced to colonies of *Apis cerana* and *mellifera*. The cerana workers removed most of the mites whereas the mellifera workers only succeeded in removing a few. It has been speculated that the mellifera workers were unable to recognise the mites as pests or their mandibles were not suitable for the job. However other studies have shown that Africanized hybrid honey bees had a mite removal rate averaging about 40%.

Strains of carniolan bees in Austria that showed some tolerance to varroasis were monitored for damaged mites. From September to April dead mites were collected daily. Over 6000 mites from 5 colonies were examined microscopically (light and scanning electron microscope). Mite damage of the legs and cuticle was evident. It appeared that the legs had been cut off leaving no frayed edges.

One particular colony was found which seemed to be quite capable of limiting the mite population to low levels. 40% of the mites had amputated limbs. With mite removal continuing through the winter months the colony seemed able to survive without any help of medicaments. There is hope that this defence mechanism is genetically controlled and can be selected for. Through natural selection such behaviour will be favoured in varroa infested colonies. Selective breeding programmes may accelerate such selection. **Ref 10**

Bees which exhibit hygienic behaviour have been shown to have a high level of tolerance to varroasis **Ref 11**. *Apis mellifera* colonies in the tropics were tested to find out if the workers could detect and remove varroa mites from sealed brood cells. The observations showed that brood infested with mites are detected by bees. The ability of bees to detect mite-infested brood cells could be influenced by the number of mites present inside the cells. The level of infestation is a variable that may influence the bees behaviour with respect to this mechanism of possible control. Uncapping and removal of brood occurred during the whole period of the study **Ref 12**. In experiments carried out in Germany Worker bees were able to detect and remove pupa infested with varroa **Ref 13**

Experiments by Christoph Otten at Mayen W. Germany have shown that there are differences in mite reproduction in European bees. Comparisons were made between:

1) Carniolan *A. mellifera carnica*
2) Italian *A. mellifera ligustica*
3) German *A. mellifera mellifera*

Rather than establishing a straight relationship between different races and tolerance to varroasis he has shown that within any one race or subspecies there are "strains" which show marked differences in their ability to "resist varroa infestations". Honey bees within a single apiary have been shown to exhibit great differences in infestation rate and varroa population dynamics **Ref 14.**

In Israel they have found differences between the resistance of Italian Caucasian and Carniolan bees. Work has already begun in selecting varroa tolerant strains. Work is also progressing privately in France and Italy.

Yugoslavian beekeepers have been selecting for varroa tolerance since 1984. Some of these bees were brought to the USA in 1989 **Ref 15** These bees seem to have a reliable resistance to varroa and acarine. Although this *A. m. carnica* subspecies is twice as resistant to Varroa as domestic bees colonies may still need chemical treatment to control severe outbreaks.

The Washington Department of Agriculture plan to release these bees, from Yugoslavia, that show some resistance to the mite. This is the first time that this agency has ever released an insect as breeding stock. The Honeybee breeding Genetics and Physiology Laboratory at Baton Rouge, plan initially to release the bees only to selected bee breeders.

Beekeepers here must be alert to any colonies that survive well when all about are losing their bees and dying from varroa. Such stocks could provide the raw material for selecting bees which have some resistance to the mites. Of course the varroa mite will also be adapting during this period. The parasite will develop an answer to any resistance selected for naturally or bred for. If varroa is like any other parasite then it would show evolutionary adaptations to survive any protective mechanism or behaviour that the bee develops. It should not be thought that breeding will ever allow varroa to be eliminated.

Ref 1 Rath, W,. (1992) The key to varroa: The drones of *Apis cerana* and their cell cap. American Bee Journal May: 329-331
Ref 2 Hanel, H. Koeniger, N. (1986). Possible regulation of the reproduction of the honeybee mite *Varroa jacobsoni* by a host's hormone: juvenile hormone III. Journal of Insect Physiology 32: 791-798
Ref 3 De Jong, D.; Goncalves, L.S.; Morse, R.A. (1984) Dependence on climate of the virulence of *Varroa jacobsoni*: Bee World. 65: 117-121
Ref 4 Smith, F. G.; (1958) Beekeeping observations in Tanganyika. Bee World 39: 29-36 Smith, F. G.; (1961) The races of honeybees in Africa. Bee World 42: 255-260. Tribe, G. D.; Fletcher, D. J. C. (1977) Rate of development of the workers of *Apis mellifera adansonii*. Pp 115-119 from African bees: taxonomy, biology and economical use. ed. D. J. C. Fletcher. Pretoria, South Africa: Apimondia.
Ref 5 Moritz, R.F.A., Mautz, D. (1990) Development of *Varroa jacobsoni* in colonies of *Apis mellifera capensis* and *Apis mellifera carnica*. Apidologie 21: 53-58.
Ref 6 Moitz, R.A.. (1985) Heritability of the post capping stage in *Apis mellifera* and its relation to varroasis resistance. J. Hered. 76: 267-270.

Ref 7 Buchler. R, Drescher W, Variance and heritability of the capped development stage in European *Apis mellifera* and its correlation with increased *Varroa jacobsoni* infestation. J.A.R. 29 (3): 1720176 (1990)

Ref 8 Mailbox (1991) American Bee Journal Jan: pp 11

Ref 9 Moritz, R.F.A. & H. Hanel. (1984) Restricted development of the parasitic mite *Varroa jacobsoni* Oud in the Cape honeybee *Apis mellifera capensis*. Esch. Zeitschrift fur Angewandte Entomologie 97: 91-5

Ref 10 Ruttner, F. & Hanel, H., 1992. Active defence against varroa mites in a carniolan strain of honeybee, Apidologie 23: 173-187

Ref 11 Taber, S. (1992) Resistant bees. Gleanings in Bee Culture: Feb. 78-79

Ref 12 Boecking O., Rath W., Drescher W., 1992. *Apis mellifera* removes *Varroa jacobsoni* and Tropilaelaps clareae from sealed brood cells in the tropics. American Bee Journal November 1992 732-734

Ref 13 Boecking, O.; Drescher,W. (1991) Response of A. *mellifera* colonies infested with varroa. Apidologie: 237-241.

Boecking, O.; Drescher,W. (1992) The removal response of A. *mellifera* colonies to brood in wax and plastic cells after artificial and natural infestation with varroa and freeze-killed brood. Experimental and Applied Acaralogy.

Ref 14 Moosbeckhofer. M., Fabsicz. M., Kahlich. A. (1988) Investigations on the correlation between rate of reproduction of *Varroa jacobsoni* and infestation rate of honeybee colonies. Apidologie 19 (2): 200-206

Ref 15 Rinderer et al. (1993) The breeding, importing, testing and general characteristics of Yugoslavian honeybees bred for resistance to

3.12 RESEARCH

National Beekeeping Unit

When the National Beekeeping Unit was threatened with closure the BBKA were requested to ask the membership to make a donation towards its costs to demonstrate that beekeepers appreciated this government service. Eventually the Annual Delegates Meeting voted to ask each member to make a donation of £1.50 for a period of three years. The offer of this money was crucial to the continuation of the N.B.U. at that time. Since then the goal posts have moved and the NBU is no longer part of MAFF it is now under the umbrella of the Central Science Laboratory. Its future now seems relatively secure. Eventually the NBU will be based near York in purpose built buildings on a new site with many other CSL departments.

The NBU is now in a position to earn money by doing research, arranging courses or carrying out trials or other investigations. In 1992 The BBKA were able to approach the NBU with a proposition that they carry out some research into the different methods of detecting varroa. In addition information was requested on the mite numbers detected and the absolute number of mites in an infested colony at any one time. Such work has been carried out elsewhere but the results are still confusing. No one really knows what the population dynamics will be in the UK. Such research will be needed in working out the threshold levels at which various treatment regimes are necessary.

The amount of money that the BBKA was able to collect and subscribe to this project was well short of the full costs but the ministry were so impressed by this commitment that they decided to divert additional funds to the project. To date the ministry's contribution is 10 times what the beekeepers have given. I would hope that the beekeepers who have handed over their £1.50s over the last three years are pleased with what their money has been able to achieve.

A research worker, Dr S.J. Martin BSc MSc Phd FRES, has been engaged to work on this project for $1^{1}/_{4}$ years. Dr Martin is not a beekeeper "yet" but his qualifications are impressive. His Honours degree (Zoology) was gained at the University of Wales in 1983. He then went on to gain a Masters degree at Shinshu University, Japan in 1987, followed by a Doctors degree back at the University of Wales in 1990. Dr Martin has undertaken scientific work in China, Japan, S.E. Asia and the Falklands. He has gained a number of awards and has 20 scientific publications to his credit. He has carried out research work with wasps, hornets, moths, beetles and bees. He has worked with *Apis cerana* beekeepers in S. Japan and was a Scientific Member of the 1992 British China Expedition during which he made a collection of bumble bees for the British Museum.

Starting in the spring of 1993 Dr Martin will be assessing varroa mite populations in infested test apiaries. The approach will be to establish a reliable methodology using acaricides to knock down the mites. The work will be replicated in different apiaries with colonies having different levels of infestation. Absolute mite levels will be determined at intervals by liquidating random colonies and counting all the mites on the bees and in the combs.

Project Summery

This project aims to provide diagnostic and monitoring tools for varroasis for the British beekeeping industry to use itself. Field diagnosis for the presence of the varroa mite is usually imprecise and the use of chemical agents, such as Bayvarol has never been quantified. This study should provide beekeepers much needed information on the relative value of the various available diagnostic aids and provide valuable essential information on mite population biology in UK conditions that will also enable beekeepers to assess and adjust treatment regimes to suit changing levels of local infestation.

Clive de Bruyn

Central Science Laboratory

Throughout the active season it is also intended to make a daily assessment of natural mite mortality to predict mite populations. At the same time the distribution of the mites in the brood and on bees will be measured. The condition of the colony and its development will also be monitored. Models will be set up to examine in detail the relationship between the stages of life of the mite on brood and adult bees to test the reliability of the data collected in the field. External influences such as weather, forage other external factors will be borne in mind.

In parallel with the above investigations certain colonies will be treated with a variety of acaricides: Bayvarol, Apistan, tobacco smoke. The mite levels in these colonies will be recorded using summer inserts and drone decapping techniques. It is hoped that this work will determine the quantitative sensitivity of these techniques in relation to seasonal changes, colony development and infestation levels.

Rothamsted

Now that varroa is present in England it will be possible for Brenda Ball to continue her investigations on viruses and varroa here in this country.

There is considerable controversy about what causes the colony to expire. It is not always colonies with the highest mite infestations that die. Curiously it has also been found that some colonies can tolerate quite high levels of mites without perishing. I am not aware that a great deal of research is being carried out elsewhere into the important part that viruses might play in varroasis.

Kashire Bee Virus has been associated by Dr Bailey with colony losses in Australia. KBV has also been found in New Zealand. Workers there claim that KBV is innocuous. Here in the British Isles we have had a ban on imports of queens from Australia because of KBV but not New Zealand. As we have been allowing N.Z. queens in now for many years KBV could already be here.

From the literature it would appear that varroa could act as a vector to spread virus diseases. New Zealand is apparently free of varroa so KBV may not be a problem. KBV has been detected in Canada but I am not aware that it is endemic. Varroa is not yet widespread in Canada. Great Britain could be the first country to have both *Varroa jacobsoni* and KBV. It might be interesting to see if these two in conjunction cause any greater problems than they do singularly.

It is fortunate that Rothamsted is still in the forefront on virus diseases of bees and we have the skill and expertise of research workers knowledgeable in this field.

In the extensive literature on varroa there is so much we still do not know. Much of the research work has been funded by chemical companies who have a vested interest in finding new products to put on the market. Such firms have the cash to dictate very often what research will be carried out.

Pure research into the biology of *Varroa jacobsoni* and the host - parasite relationship has not been covered so well. In the last decade the research carried out in the British Isles has suffered from government cuts. The pioneering work of Rothamsted under
- Dr Butler   queen substance
- Dr Simpson   swarming
- Dr Bailey   bee disease
- Dr Free   pollination

has continued with Ingrid Williams and Brenda Ball but with reduced funding.

Other institutions such as Cardiff University, Plymouth Polytechnic and Queen Mary's College have contributed to our knowledge on bees. These establishments have relied heavily on the dynamism and energy of individuals such as:
- Prof. Pickard   Cardiff
- Prof. Heath   Plymouth
- Dr Goodman   Queen Mary's.

Perhaps in the future British beekeepers are going to have to put their hands in their pockets if they want someone to do research. This is what happens in many other countries. It is noticeable that German, French and Dutch beekeepers do a significant amount in paying for research work done in their countries. This money also gives them a say in what work is carried out.

We in England have made a start. The

Clive de Bruyn

Essex resolution of some years ago to ask for 10p per member of the BBKA so that research could be carried out into European Foul Brood has now reached fruition. A C.A.S.E. award has enabled this work to be undertaken at Cardiff University.

I would hope that in the future The Bee Farmers' Association. British BKA, Welsh BKA, Scottish BA, and CONBA. would all try to raise funds from their members to carry out important research work on bees.

Progress meeting on EFB research
Left to right: Medwin Bew NBU, Brian Dancer Cardiff University, another Project worker, Prof. Pickard Cardiff University, David Little BBKA.

3.13 EPILOGUE

During the five years that I worked as an advisor at the NBU It was not considered prudent to advise beekeepers too directly what they should do with their bees. My way of getting round this restriction was to say, "This is what I would do in your case".

I have tried not to instruct anyone to adopt a particular regime of treatment as I do not know enough about varroa, its interaction with bees in this country and the efficacy and safety of the treatments advocated. However, one cannot sit on the fence forever. These are my plans for 1993.

I have been monitoring all my apiaries for varroa during the last 10 years. Winter debris has been sent off to the NBU or examined by me each year. In the last four years I have also done the tobacco test in spring and autumn. Last October all colonies were tested with Bayvarol. At this time of writing (April 1st) half my apiaries have been tested again. So far no mites have been found. Despite these searches I know for certain that I will find varroa eventually.

I am making varroa screens for that third which do not have them at the moment. These will be left permanently on the hive throughout the year. This is a nuisance as all my floors are stapled permanently to the brood box as most of the colonies are moved during the year. Varroa screens are going to be one more piece of equipment to lug around. The inserts will be regularly examined at each routine inspection. I feel quite confident that I can spot varroa now that I know what to look for. I might also try uncapping drone brood and sampling adult bees merely for the experience. I have never found any beekeeping operation as easy as the books say so I might as well sort out the wrinkles now before things get really hectic.

All swarms collected will be checked with Bayvarol. I will be looking in on my neighbouring beekeepers to see what they are doing. Hopefully they will be aware and looking. Where they are not perhaps I can tactfully encourage them to monitor their hives. If the beekeeper is genuinely unable to cope I might consider doing some checking myself.

Now what happens when I find varroa? If it is in a swarm and I have found no varroa in any of my apiaries I would destroy the bees. Once I have an infested apiary then that is where infested swarms will end up. On discovering varroa in an apiary I would have to inform the ministry. After that I will be on my own. I would attempt to assess the level of infestation. I know that this is not easy as the population dynamics of the mites varies throughout

Clive de Bruyn

the year. If the discovery is early in the year I would see what effect drone trapping will have. Presumably the infestation would not be heavy and I might be able to keep mite numbers at bay for some time with bio-technical methods.

I would try to avoid moving colonies which I know to be infested to other "clean" sites in the county or outside. As the infestation spreads through the natural movement of infested bees or by migratory beekeepers all my apiaries will have varroa. Until that time I will not intentionally spread known infested colonies to "clean" sites. As normally I move bees to pollinate and migrate to crops in the county and outside this will place some restrictions on my beekeeping. Where other beekeepers and I share sites it will be important for me to know if they are looking for varroa and whether they have found any mites.

If I am lucky and my colonies appear free of the mite I will try to gain some experience in some of the biotechnical techniques available. There is apparatus to be tried and tested. there are techniques to be worked out. Given sufficient time I will experiment with some of the soft chemicals and see what effect dusts and volatile oils have on the colony. I have no doubt that I will get feedback from the bees as to whether they approve or not. I can then see if I can adapt the technique so that it suits me and the bees.

I will have to work out a management system that will suit my beekeeping. I will have to integrate varroa control with my queen rearing and swarm control. I need to develop a method of raising drones without losing them to varroa. As a person who likes to believe that he can practise selective queen rearing I think that my drones are the most valuable asset I have. How can I carry on producing sufficient virile drones from varroa infested colonies? From the limited work some of us are doing already towards providing drones for A.I. it may be necessary to produce drones from strong colonies where an acaricide is used to limit the damage from varroa.

These colonies I would not use for honey production.

It might be that my attempts at breeding could benefit eventually. If I can arrange to produce drones when all around me are loosing theirs to varroa then I should stand a higher chance of my virgin queens mating to my drones rather than anyone else's. I do not know yet what the situation will be with queenless nuclei used for mating. I suspect that these small units which are already under considerable stress will probably need some extra management. Acaricide treatment may be necessary.

When eventually my mite numbers justify chemical treatment I will carefully consider the moral and ethical implications. I am probably like many other beekeepers. I have too many colonies. I built up my stocks in the bad years to give me sufficient honey.

It will be necessary to try and keep abreast with research being done here and abroad. I must try and obtain the original research papers wherever possible. It is only when one goes back to the original work that the tenuous data on which all future speculation is based can be appreciated. To quote Professor Heath, "It ain't what you know that lets you down but what you think you know that aint so". Once a statement is made it gains credibility as everyone else refers to it in subsequent papers. The time taken for the varroa mite to develop from egg to mature adult and the sex of the eggs laid is an example. My co-author has gone into detail over this. There is now some doubt on the original work which claimed that Asian mites in an observation hive could remove over 90% of the varroa mites introduced. It appears that the mites used were taken from mellifera bees. Perhaps the cerana bees reacted to the "mellifera smell" on the mites? **Ref 1** I believe that in nearly all instances the scientists are accurate in what they report, it is the interpretations that others put into what they report that can lead to confusion.

Ref 1 Boecking, O., Rath W,. & Drescher W,. (1993): Grooming and Removal Behavior - Strategies of *Apis mellifera* and *Apis cerana* bees against *Varroa jacobsoni* American Bee Journal Feb 93 pp 117-119